T0205505

Mine Wastes and Water, Ecological Engineering and Metals Extraction

Margarete Kalin-Seidenfaden
William N. Wheeler

Editors

Mine Wastes and Water, Ecological Engineering and Metals Extraction

Sustainability and Circular Economy

 Springer

Editors
Margarete Kalin-Seidenfaden (iD)
Boojum Research Ltd.
Toronto, ON, Canada

William N. Wheeler
Boojum Research Ltd.
Toronto, ON, Canada

ISBN 978-3-030-84653-4   ISBN 978-3-030-84651-0   (eBook)
https://doi.org/10.1007/978-3-030-84651-0

This Springer imprint is published by the registered company Springer Nature Switzerland AG
The registered company address is: Gewerbestrasse 11, 6330 Cham, Switzerland

I would like to dedicate this book to the late Jui-Lin Yen (Allen), my long-time partner, and to Patricia (Pat) Sudbury. Without Allen's wisdom and support, I would not have been able to carry on my quest for the ecology of mine waste and water. Pat listened patiently to the many endless discussions between Mike and myself for many years, always being provided with nice lunches prepared by her. She kindly accepted many times, and even during the retirement of her husband, to stay alone. Mike and I visited so many waste sites as he wanted to see for himself the ecological wonders which I had reported to him.

May, 26th 2021
Margarete Kalin-Seidenfaden Hasliberg-Reuti, Switzerland

Foreword: Viewpoint on the Founder

I met Margarete Kalin many years ago because she is the daughter of a war comrade of my father-in-law. Since the 1970s, when Margarete had emigrated to Canada, the contact became more intensive. She visited us frequently at that time in connection with international professional conferences around the world. In the meantime, she had become an associate professor and, after several decades, was working with her scientific team on sustainable methods for treating mining legacies.

Thus, I was introduced to a world previously unknown to me, mining, and its wastes. They are festering wounds in the surface of the earth worldwide. As a businessman, it quickly became clear to me that these un-remediated contaminated sites from the extraction of coal, copper, or uranium, for example, will continue to burden companies, states, and governments economically and ecologically for many generations to come.

Their wastewater, in particular, is being released in ever-increasing quantities through precipitation and enriched by weathered sediments. Particularly noticeable is the increasing clogging by precipitated rust-brown iron hydroxide in (artificial) lakes, streams, and rivers. Such turbid water is not inviting for swimming or recreation. Nor as a habitat for plant and wildlife let alone for consumption as (treated) drinking water.

Margarete conducted extensive studies and experiments on many "lunar landscapes" around the world. New or adapted methods of innovative biological and microbiological processes were often used. The series of experiments and practical results were meticulously documented in electronic databases and constantly adapted to current IT procedures. In the meantime, the library has a volume of about 160 documents.

Margarete has learned from childhood to adapt her life and her research work to tough environmental conditions and to persevere. As the daughter of a regime-critical Protestant pastor and a teacher, she lived with her two siblings in Thuringia East Germany from 1947 to 1960 under constant observation by state security. When the political pressure became too great, the parents decided to flee to the West with their three minor children, and they finally ended up in the Bernese Oberland Switzerland, where the father got a job as pastor of a parish.

During his first mountain climb, the father died unexpectedly of a stroke and the mother had to support the family as a teacher in the remote mountain village. Since there was not much else to distract her except skiing in winter and hiking in summer, the young Margarete became intensively involved in observing the rugged nature of the mountain world, thus laying the foundation for her later studies and scientific work.

In 1971, the trained executive secretary and agricultural-biotechnical laboratory technician emigrated to Canada with her husband. There she worked at McGill University as a dishwasher, but her undercover was soon detected. She complained to the boss about the radioactive filter papers in the wastepaper bucket beside her fume hood. While living in Zurich applied to three universities as a "mature student" for a bachelor's degree. She typed her husband's doctoral thesis in the evenings entering the final corrections. After her apprenticeship, Margarete became a lab assistant at ETH in Zurich. During that time, she attended courses that interested her, including working on experiments involving the genes of fruit flies. She attended lectures of her interest during the day and worked on experiments with fruit fly genetics in the evenings to replace lecture time.

Of the three universities, one refused her admission on the grounds that her father had been an East German spy: deeply frightening news for Margarete. But the other two universities gave the green light, and after five years, both the Bachelor of Science and Masters in Entomology at the University of Toronto were completed. The documentation on the social behavior of insect larvae was even published in *Nature*, the quintessential journal of science.

This unusual distinction for a nonacademic (without a doctorate) helped her to obtain a position as a research associate at the Institute for Environmental Studies (IES) at the University of Toronto. She was given the task of conducting research on the ecological behavior of the waste piles of a uranium mine north of Toronto. A final report requires that literature in the area must be cited in every report. Margarete found that the university's libraries had no scientific literature on plants, trees, and shrubs or grass growing on uranium wastes. This prompted her to propose a four-year study to the government of Canada, for which she prepared a research plan to determine the possibility of radioactive contamination of the food chain.

This project was approved by the Canadian federal government, and she investigated uranium tailings in Ontario and soon all uranium tailings sites in Canada. This was the first time she entered the tough male domain of mining. It was certainly not an easy learning curve, but she always kept her head up. Through the publications of these studies, Margarete was appointed as Canadian representative to the International Atomic Energy Agency (IAEA), much to the surprise and envy of her colleagues at the institute, who each had a doctorate.

Working on projects in the uranium industry soon attracted the interest of other mining companies, and it was natural to start her own company, Boojum Research Limited, in 1982. Because of Margaret's background, Boojum's approach has always been one of practicality and innovative problem solving, often in the face of opposition from established consultancies and mining professionals. This led to another milestone in 1989, as her work on a closed copper zinc mine was published in the first edition of the book on *Ecological Engineering: An Introduction to Ecotechnology*. Over the years, she earned increasing respect being awarded prizes by the CLRA (Canadian Land Reclamation Association) as well as the CIM (Canadian Institute of Mining, Metallurgy and Petroleum).

What does this mean for the benefits and practical application in ecology, economy, and politics? Here are some examples from East Germany: Until the reunification of Germany, large areas in the former GDR were characterized by the key industries of lignite, potash, spar, and ore mining and processing as well as uranium mining. All sectors produced enormous amounts of overburden and landfill material with well-known negative consequences for the environment, especially water pollution. Visible traces of the contaminated sites from coal mines can currently also be found in the river Spree which flows through Berlin. Nearly all tributaries are also affected. The result is that landscapes are covered by washed-out iron hydroxide from the buried coal, mainly in Saxony, Saxony-Anhalt, Brandenburg, and Thuringia.

The Wismut GmbH (uranium for Russia) was founded in 1991 after the Berlin wall fell and LMBV (Lausitzer und Mitteldeutscher Bergbau-Verwaltungsgesellschaft GmbH coal) in 1995 to manage, rehabilitate, and prepare for subsequent use. Both companies are federally owned and are under the control of the Federal Ministry for Economic Affairs and Energy and the Federal Ministry of Finance, respectively, and will continue to devote themselves to the perpetual burdens for a long time to come, critically eyed by politicians, environmentalists, and citizens' initiatives, such as the "Klare Spree" action alliance.

In 2005, the LMBV organized the first International Remediation Congress ISC in Berlin, in which Margarete also participated as a keynote speaker. Wismut in turn invited Margarete to give lectures and workshops to report on her experiences with Ecological Engineering processes. A wetland treatment system was changed into an ecological engineering system. The system showed great interest and worked reasonably well however missed some essential aspects, as her participation was terminated too early. Their own conventional processes took over. Margarete is convinced that there is still considerable potential for innovation, rationalization, and thus cost reduction.

In order to make her scientific documents and evaluations available to the general public for research and teaching as well as for operational use and to establish them as important building blocks in her fields of expertise, Margarete donated a truckload of books to the mining and environment library at the Laurentian University of Sudbury in Canada, supporting with a stipend the integration into the online library. To round it off, she also offers workshops and lectures.

For her "life's work," she was awarded an honorary doctorate from this university on October 2020. Her long-time co-author who became her, friend made great scientific contributions. Dr. Günther Meinrath from Passau and Bad Tölz commented on his laudation with the words: "The University of Sudbury does not only distinguish Margarete Kalin-Seidenfaden. The Laurentian University of Sudbury also distinguishes itself – with one of the most unusual personalities the scientific world of today has to offer."

Gustav A. Habenstein, Chief Representative. Always Solutions Services GmbH. Translated with www.DeepL.com/Translator (free version)

Always Solutions Services Gmbh   Gustav Habenstein
Moers, Nordrhein-Westfalen, Germany

Foreword: Frame of Reference on the Early Projects

I was asked by Margarete Kalin to write some introductory comments for this ecological engineering book, providing some thoughts on this topic as viewed from the mining sector perspective. Specifically, my viewpoint relates to the uranium mining sector as carried out in northern Saskatchewan. I worked most of my career for Cameco Corporation, covering both its Fuel Services and Mining sectors. Having retired a decade ago, I can only provide my personal perspective.

The uranium mining sector in northern Saskatchewan, which while not large in scale relative to other mining sectors, is in my view a good example of how a well-regulated, responsibly operated mining sector can successfully manage its environmental impacts, both now and in the future. There has been clear progress in dealing with past issues, as well as implementing and improving current environmental performance. I worked with Margarete on several Boojum projects with Cameco, involving both historic and current operating uranium mine/mill environmental matters. I should first summarize why we chose to fund the Boojum research work described in this book.

Our objective centres on continually improving the understanding of how our mining operations interact with the local environment. This interaction primarily focuses on the water environment with topics relating to how tailings, waste rock and overburden stockpiles are managed, as well as the behaviour of former mine pits and underground workings. Uranium is a relatively common naturally occurring high-value metal. It is found throughout the world in low concentrations, with known pathways of dispersion and concentration through chemical and biological processes. One can learn a lot about uranium's environmental behavior from how it has been naturally distributed around viable ore bodies. For instance, this type of knowledge can lead to the discovery of mineable deposits by analysing such things as downstream lake bottom sediments and past glacier transport activity.

In Saskatchewan, to licence and operate a modern uranium mine, credible decommissioning plans must be developed from the outset. Such plans must be supported with justified cost estimates, then approved by regulatory authorities and typically financially backed up by acceptable financial assurance mechanisms. These financial assurances are required if the operator defaults on their decommissioning obligations. Decommissioning plans must be supported by long-term environmental risk assessments, largely built on environmental transport models. Source terms for any surface and groundwater contaminant transport pathways to the receiving environment are needed, including such things as expected adjustments with time, as well as contaminant transport dispersion and accumulation characteristics. Good models require good input data, based on knowledge of how all contaminants of concern behave, spanning both radioactive and non-radioactive constituents. In northern Saskatchewan mines, in addition to uranium and its decay products, such as radium, other key constituents include arsenic, nickel, molybdenum and selenium.

Boojum's work largely centred on the behaviour of uranium, radium-226, arsenic and nickel. The work focused on characterising contaminant source terms, processes that mobilize contaminants, processes that sequester soluble contaminants to less soluble precipitates and study of ways in which these natural processes could be optimised. The overarching decommissioning goal is to minimise the long-term mobilization of contaminants in tailings, waste rock and mined-out ore deposits. Nevertheless, some limited contaminant migration is inevitable. Reliably understood contaminant chemical characterization and interactions with the receiving environment are of paramount importance in predicting future behaviour measured over thousands of years. The halo mineralization found in displaced waste rock and ore body overburden as well as mineralization left in place in mined-out pits and underground workings must be characterized. These components may have similar mineralization to that found in the ore body itself or may reflect natural migration from the ore body. Either way, these source terms are now potentially subject to accelerated aging by being exposed to weathering factors, such as oxidation. Mill tailings, on the other hand, have completely modified mineralization due to the chemical processes used to extract the uranium.

This contaminant characterisation work continues to evolve as the mine site transitions through the various phases of its decommissioning plan. Typically, the mine site is expected to transform from an initial post-closure active management decommissioning phase to a subsequent passive decommissioning monitoring phase, before reaching the final post-licence management phase.

In addition to the primary mission of better understanding contaminant behaviours for decommissioning process approvals, ecological engineering processes can provide viable supporting remediation processes when faced with residual longer-term low-grade, low-rate contaminant transport. While long-term solutions obviously need to be based on something beyond simply saying that "nature will take care of itself", the positive aspects of naturally occurring processes to mediate any residual impacts cannot be summarily dismissed, especially in the later, more passive, stages of the decommissioning process. Reaching the fully decommissioned end-state is no simple task, but more achievable when built on good science.

CAMECO, Saskatoon   John Jarrell (Retired)
Saskatchewan, Canada

Foreword: Point of View on the Business Aspects

Progress towards long-term sustainable environmental solutions in the wake of circular economy and sustainability needs to start with a change in attitude. The executives in the mining industry should pay heed to what has happened over the course of the last few years to the automotive industry. Internal combustion engines have been declared "persona non grata" in many jurisdictions. A significant number of nations have declared that all new vehicle sales must be electric. Some have set a target of 2035, others a few years later. While the time frames may vary, the direction the automotive industry is being forced to follow is clear. These are not consumer-led initiatives. Rather, elected officials are responding to their perception of the demands of the public for environmentally sustainable policies.

I have had the privilege of knowing Margarete Kalin for more than three decades. Over the course of those years, my role has been a combination of business advisor, corporate director and friend. Her passion for creating sustainable environmental solutions for the long-term acid mine drainage problems that are endemic to mine sites has not waned. This book is another chapter in her efforts to effect change. Her firm, founded in 1982, in the first years the uranium industry, the most progressive of the industry, supported her endeavours. Projects to develop site-specific decommissioning scenarios with several large corporations followed.

I was introduced to Margarete in 1990. She was struggling to understand the decisions being made by her customers. My experience with corporations in the resource industries and understanding of general corporate behaviour helped me provide Margarete with guidance regarding the apparent lack of willingness of the mining industry to embrace sustainable solutions. Margarete refused to accept that acid mine drainage is a price the environment must pay for mineral extraction. Her perseverance has led to an impressive scientific foundation that can be used to bring an end to acid mine drainage. This book is written for the mining industry and its executives who remain firmly entrenched in the remediation approaches of the past.

Environmental restoration tools regarding the wastes as extreme ecosystems developed by Margarete Kalin with her team of scientists have demonstrated that they can meet the minimum regulatory standards by keeping contaminants within the mine waste and water management systems. Co-authors of this book are persons well known in the mining industry. They offer critical thought on land and water usage, present alternatives and encourage change which can be brought about with lower operating costs through more efficient hydrometallurgy. They support the approaches demonstrated by Boojum as the vastness of Germany's coal strip mining is indeed to anyone an extreme ecosystem. Thus, this book provides a significant added benefit if continued and further developed not only through creating jobs but also to address severe problems of the water supply in the coal-mining districts. Ecological engineering tools help gradually eliminating the perpetual liability associated with mining and their wastes.

Margarete has laid the groundwork to introduce some degree of sustainability to and a circular economy for the industry. It is now up to the executives to demand that this body of work is applied from exploration through operation and to the grave. Leaders lead…. Anyone who is not leading is following. Leaders can influence outcomes. Followers suffer the consequences. If the executives of mining companies want to have a meaningful impact on the direction that new environmental responsibility regulations take, they need to lead. If the executives do not act, governments will set new, higher minimum standards for mine site environmental compliance. The executives in the mining industry should keep in mind that whether new regulations will be effective, let alone cost efficient are seldom important criteria for government policy. The starting point rests with the executives of the mining industry. They need to change their mindset. The clock is ticking.

Boojum Research Ltd.   Ronald Benn
ON, Canada

Preface: Why We Have Written This Book

Mining, as it is currently practiced, is an extremely wasteful pursuit. A space alien watching us mine a tonne of rock – and throw 99% of it back, as we do with gold ores and nickel laterites, for example – would correctly think we are crazy. The act of mining the ore is the most expensive part of metals extraction, yet we happily throw most of what we mine away, at the same time creating vast amounts of tailings, waste rock and contaminated fresh water. The mining industry needs a paradigm shift in thinking to recognise that supposedly uneconomic materials that we currently throw away – such as iron, magnesium, silica, calcium and aluminium – are finite and valuable resources in themselves. We have taken the trouble to dig them up, so we should take further trouble to recover them. This is the first reason for writing the book.

The book is targeted at a wide-spectrum audience, mining executives and professionals, government regulators, environmentalists and anyone with an interest in our planet and its ecosystems. However, it is the mining executives and government that we would like to embrace the Way Forward, because stewardship of our planet is a necessity. Current practices must change. Society, in general, recognises that there is a crisis with climate change and global warming. Mitigation of mining wastes is thought to be being driven by the so-called electric revolution. This revolution needs metals and minerals in large amounts to facilitate the dream of renewable energy. The waste sites of these minerals are growing, consuming more land and contaminating more water. The second reason for writing the book is to offer suggestions of ways to responsibly go about mining.

The book also describes the balance between mining and ecology, and how the latter, particularly the roles of microbes, can assist in mitigating, reversing and containing much of the damage that mining has created. It is a topic that is not generally well understood, and to which there is a general reluctance to embrace, but it is one that we believe is crucial and essential to preserve the viability of the planet. This is the third reason for writing the book. Contaminated drainage from mining wastes, as it is believed, is not "the price to be paid for metals".

Margarete Kalin-Seidenfaden earned the nicknames "Cattail Kalin" and "Swamp Doctor" for her professional life as a lone pioneer. She was initially tolerated and subsequently respected by the industry, culminating with the award of a PhD Honoris Causa from Laurentian University, based on her unique experience of investigating and working on waste piles and their drainage, first in Canada and its Arctic, then in Germany, China, Australia, Brazil and the USA. Different commodities were encountered with the work, but the same extreme ecosystem ecology was always present. This supports the claim that ecological engineering processes have global relevance and need to be further developed and expanded, based on biogeochemistry and microbial ecology. The industry was interested in Boojum's work, the forerunner being the uranium industry and other commodities followed, but after three decades the interest dwindled. We all know that change is difficult, but time is now pressing, which is the fourth reason for writing this book.

The "Blue Marble", planet Earth, is our home, the only one we have. Unlike our forefathers who were able to drop everything and sail off to the New World, we are not able to do that. Astronomers estimate that there are over two billion possibly habitable exo-planets in the universe, but the nearest of these is over four light years away. In the far distant future, humankind may be able to reach out to the stars, but for now we are confined to our planet, so we need to look after it.

The Earth is a truly marvellous, wonderful ecosystem, but it is a fragile one, and we are duty-bound, if we want to continue to exist here, to show appropriate stewardship. This is really the first and main reason for writing this book. Although a cliché, it is nevertheless true that we either mine or grow everything we use. The products of the former underpin our modern society, but they are not renewable, they are finite. Only agriculture is truly renewable. Freshwater is somewhere in between since it is recycled on a global scale. Humankind has recognised the use of metals since antiquity, but it is only in the last few centuries or so that large-scale mining has created numerous new environmental issues. Freshwater is also a finite, but recyclable resource and needed for mining and mineral extraction. Desalination can go some way to alleviating the pressures on freshwater use, but it is not the answer, being both costly and returning the concentrated brine creates additional problems for already-distressed oceans.

Beyond the obvious environmental issues, therefore, are population growth, and the use of arable land and freshwater. These reasons are highlighted in our book and we hope that they are sufficient arguments to provoke action. It is also our hope that in writing this book, we have contributed not only to the general discussion on this topic, but we have also presented some concrete ideas on how improved stewardship of our home can be implemented.

Oakville, Canada   Michael P. Sudbury

Alexandria, Canada   Bryn Harris

Toronto, Canada   Margarete Kalin-Seidenfaden

Acknowledgments

The ecological fieldwork on the mine waste management areas was only possible because many mine managers, too numerous to name, were instrumental in giving access to the sites and actively supporting our work. Boojum is grateful for the first hand knowledge about mine operations generating the waste rock piles and the processes which reject barren liquors and tailings. These two ingredients were necessary to make progress possible. Without the acceptance of the Canadian Mineral Processors (CMP'ers), this work, developing the Ecological Engineering tools, could not have been realized. Our gratitude goes to all the scientists in geology, hydrology, mineralogy and microbial ecology, algal physiology, data management and not to forget the many summer students who needed often courage when requested to carry out nearly impossible tasks. These waste sites are no picnic places. Many thanks to our coauthors, Michael P. Sudbury and Bryn Harris, who provided a framework for the Ecological Engineering tools as their work support solutions to the way forward.

Toronto, Canada   Margarete Kalin-Seidenfaden

Toronto, Canada   William N. Wheeler

Summary

This book, in ten chapters, addresses and presents methods to stabilize mine waste and water management areas. It presents scientific bases to utilize ecological processes in order to balance the weathering processes in these extreme ecosystems with a realistic view of these waste sites of broken or ground rock. It has been the self-inflicted mission of Boojum Research Ltd., supported firstly by the uranium industry, then followed by several others to develop decommissioning scenarios. Over 40 years, it grew into an approach, which if further developed with the mining industry, may lead to a "Eureka moment" and may make mine waste management possible from "the cradle to the grave," given that one of the most relevant findings is that "bugs fight bugs." The book is intended to be the beginning of an approach which might lead to the "holy grail" long awaited by the industry, consultancies, and academia.

The importance of the mining industry is well-known, if not always appreciated, as our lives are surrounded by its products, i.e., metals and industrial minerals. As long ago as 1556, Georgius Agricola first highlighted the destructive environmental side-effects of mining and metals extraction, namely dead fish and poisoned water. These effects, unfortunately, are still with us today. Since then, our knowledge of the reasons leading to environmental deterioration have grown tremendously, and many mining and smelting processes have been improved. In Agricola's time, mining was generally local and surrounded by forests which were cut to support the underground tunnels of the mine and used for fuel for roasting of the ores. However, changes in mining methods since the early 1900s to open pit mines, with or without tunnels, became the most common mining method. This led to a second mining waste namely, vast rock piles with uneconomical concentrations of mineral from the overburden, in addition to the ground rock from the mills, namely the tailings. Along with population growth and technology advancements, metal demand increased, and further wastes were created due to chemical neutralization of mine wastewater, generating a mineral-laden sludge. This latter material, however, is in fact just a perfect alternate food for microbes.

An umbrella large enough and/or the best cover strong enough to persist for decades or millennia for enclosing the wastes does not exist. To halt weathering, the oxidative reaction expedited by microbial activity has to be restrained or even inhibited. The logical approach to address this is to reduce the weathering rate and promote precipitation within the waste deposits by delivering, with the oxygenated water, a reactant which alters the mineral surface such that oxygen-consuming microbes dominate.

Boojum was guided by professionals of the mining industry from its inception in 1982 in Toronto, Ontario, Canada, and continuing up to the idea of summarizing about 40 plus years of Ecological Engineering methods in a book. A partial summary is available in the Oxford Research Encyclopedia under the title Mining, Ecological Engineering and Mineral Extraction for the twenty-first century by Kalin et al. (2018). Chapters 1, 2, and 3 are included in this book with an enlarged Chapter 4 and generally more data have been integrated. The text is expanded in the hope that Ecological Engineering tools can be further developed and applied.

Chapter 1 in this book describes mining waste generation, weathering, and the complexities of predicting the characteristics of their effluents from the onset of operation. Chapter 2 provides rough estimates of global annual ore and waste generation of major commodities, their water, and land use reflecting the urgent need for action to change the present approaches to mining and waste generation. Chapter 3 provides a solution to both, giving an outline of changes needed in hydrometallurgical processes to recover more commodities than the target mineral, and also saving land and water. Chapter 4 presents a brief history of mine waste management funded by various government and industry programs, with Boojum reviewing its early contracts. The prevailing response was that ecological engineering and biology do not work in the winter, and certainly not in the far North of Canada. This misperception provided a unique opportunity to document natural recovery processes of long-time abandoned operations in the North West Territories and the Yukon. It was founded by Canada's DIAND (Department of Indian Affairs and Northern Development) and facilitated by the water boards of both jurisdictions, providing water monitoring data and aerial photographs which were used to assess the recovery of the land disturbance. Out of 31 sites investigated, only 5 showed no recovery of the water quality (mostly small gold mines where the gold was collected with mercury). With these findings of nature's repair potential, Boojum gained confidence, and despite Boojum's initially cautious faith, these natural cleansing processes were confirmed. Yes, they take a bit of time – but they work and are sustainable.

In Chap. 5, the differences in degradation of organic and inorganic pollutants are highlighted. The use of natural or constructed wetlands with sediments within which plants grow is very successful cleansing and degrading agents for organics. However, when tested for mining wastewater containing inorganic contaminants, the treatment generated hydrological problems. The different removal processes are described.

Chapters 6, 7 and 8 describe how the ecological engineering tools which are needed to neutralize acid mine and rock drainage. These tools, namely ARUM (Acid Reduction Using Microbiology), biological polishing and biofilm augmentation to reduce sulfide oxidation, are installed within the mine waste management area. As ditches and shallow ponds are void of sediments those have to be constructed. Instructions are given to provide conditions for microbially active sediments.

In order to remove contaminants out of large water bodies, lakes or flooded pits (also called pit lakes), it is necessary to generate particle forming organics, phytoplankton, through adjusting the ratio of nutrients in the water. These free-floating algae provide, in the water column, cell wall surfaces to which metals adhere/absorb, forming particles large and heavy enough to settle to the bottom sediment, again either constructed or existing. For attached periphyton growth (attached living algae), surfaces have to be provided supported by floats or adding brush at a depth where light can penetrate, at the edges of pits or pools, initiating the growth of a floating living cover. These measures have in common that the lack of a continued supply of organics halts a sustainable treatment approach within the water body. This tool is referred to as biological polishing, replacing the use of flocculation agents for particle formation (Chap. 7).

Lastly, a very important tool is the one presented in Chap. 8. To various acid-generating mining wastes, Carbonaceous Phosphate Mining Wastes (CPMW) were added, weathering products/particulates of CPMW were carried with the rain to the mineral surface, the effluents had a circumneutral pH. This process was brought about through the formation of a biofilm over the mineral surface and has been documented several times, by different scientific groups repeating tests with different commodities and experimental designs. Initial skepticism suggested that biofilms would not last!! The rocks from the first experiment have been stored for 11 years, as Boojum anticipated the biofilm might no last, but sometime every experiment has an end. The rocks were placed outdoors without further CPMW addition for an additional two full years, with continued improvement of the effluent. Hence, outdoor exposure was continued, and eventually, the surfaces were investigated through SEM microscopy at the University of Toronto. Eleven years had passed, and some of the biofilms still persisted. Recently, Boojum found in updating the literature on MIC (Microbial Inhibition of Corrosion), a publication about a rust-free nail covered by a biofilm several thousand years old.

Chapter 9 highlights R&D projects that have used the above tools to contain or slow the production of AMD in mine waste management. Chapter 10 provides some practical suggestions to move the industry closer to sustainability and some of the Sustainable Development Goals of the United Nations.

Margarete Kalin-Seidenfaden

William N. Wheeler

Contents

1   **Introduction and Weathering**.   1
    Margarete Kalin-Seidenfaden

2   **Dimensions of Global Mining Waste Generation and Water Use**. . . .   9
    Michael P. Sudbury

3   **Toward a Sustainable Metals Extraction Technology**   17
    Bryn Harris

4   **Waste Management: A Brief History and the Present State**.   29
    Margarete Kalin-Seidenfaden

5   **Constructed Wetlands and the Ecology of Extreme Ecosystems**   41
    Margarete Kalin-Seidenfaden

6   **Ecological Engineering Tools in Extreme Ecosystems**.   47
    Margarete Kalin-Seidenfaden

7   **Biological Polishing Tool: Element Removal
    in the Water Column**.   73
    William N. Wheeler, Carlos Paulo, Anne Herbst, Hendrik Schubert,
    Guenther Meinrath, and Margarete Kalin-Seidenfaden

8   **The Biofilm Generation Tool for the Reduction
    of Sulfate Oxidation**   105
    Margarete Kalin-Seidenfaden

9   **R&D Field Applications**   121
    Margarete Kalin-Seidenfaden

10 **The Way Forward**   147
    Margarete Kalin-Seidenfaden, Michael P. Sudbury, and Bryn Harris

Related Reading   151

Contributors

Bryn Harris Alexandria, ON, Canada

Anne Herbst Department Maritime Systeme, Interdisziplinäre Fakultät, Rostock, Germany

Margarete Kalin-Seidenfaden Boojum Research Ltd., Toronto, ON, Canada

Guenther Meinrath Head, RER Consultants Passau, Passau, Bavaria, Germany

Carlos Paulo School of the Environment, Trent University, Peterborough, ON, Canada

Hendrik Schubert Universität Rostock, Biowissenschaften Lehrstuhl für Ökologie, Rostock, Germany

Michael P. Sudbury Michael P. Sudbury Consulting Services Inc., Oakville, ON, Canada

William N. Wheeler Boojum Research Ltd., Toronto, ON, Canada

Chapter 1
Introduction and Weathering

Margarete Kalin-Seidenfaden (iD)

Abstract This chapter introduces mining wastes, primarily from sulfidic ores. These wastes are environmentally destructive and have longevities of thousands of years. The root cause of mining waste is the weathering of exposed waste rock and ores. The weathering process is exacerbated by microbial metabolism. The focus of this chapter and book is the delineation of the role of oxidizing microbes in causing mine waste effluents and the role of reducing microbes in their prevention and control.

Weathering liberates minerals, too low in concentration to be milled economically. The surface area within mine waste is dramatically increased and therefore accessible for oxidation, e.g., weathering. Many elements liberated are needed to sustain life, but increased concentrations are toxic when reaching the receiving environments. The chapter gives a brief, but informative overview of mining practices, and about the complex factors which contribute to the rate and extent of weathering processes. Challenges in predicting the weathering products such as acid mine or rock drainage are presented.

Keywords Mining wastes · Mining technology · Mineral processing · Extractive metallurgy · Environmental liability · Waste mineralization · Weathering processes · Challenges in projection of effluent characteristics · Acid-Base-accounting,

Mining and the extraction of metals have been a large part of human activity since prehistoric times, and the modern world has an almost unquenchable thirst for more and more metals, not only the traditional metals, such as iron, copper, and aluminum, but also the rare and so-called rare-earth metals, which are important for mobile phones and the like. However, recovering metals has not come without a cost. The environmental consequences of even ancient activity are still evident

M. Kalin-Seidenfaden (✉)
Boojum Research Ltd., Toronto, ON, Canada
e-mail: margarete.kalin@utoronto.ca

M. Kalin-Seidenfaden, W. N. Wheeler (eds.), *Mine Wastes and Water,
Ecological Engineering and Metals Extraction*,
https://doi.org/10.1007/978-3-030-84651-0_1

today. For example, there is still evidence in Spain of mining by the Iberians in 3000 BCE (Davis et al., 2000), in the old mines of the Incas in South America (Strosnider et al., 2011), and more recently, in the vast "red mud" ponds from bauxite mining found scattered worldwide (Ritter, 2014). There is, perhaps, no greater and more poignant reminder of this latter environmental legacy than the fact that October 2016 was the 50th anniversary of the terrible Aberfan disaster in Wales, where an unstable coal mining tip engulfed a school, killing 116 children and 28 adults. More recently, the tailings dam break in Brazil is another example of a mining tragedy that cost 248 lives and considerable environmental destruction (Wise Uranium Project, 2019).

The practice of mining and metals extraction is that the mined, broken mineral is separated into rocks that do not contain sufficient metal to be economically extracted, referred to as waste rock, and rocks that contain ore. These latter rocks are ground, the ore is extracted, and the remaining ground rock disposed of as tailings. The wastes generate effluent known as mine or rock drainage, which is either alkaline or acidic. Since ancient times, mining methods and practices have, of course, changed dramatically, currently having a much larger scale and much more mechanized, and with commensurately greater environmental issues. Ore was once high-graded (collected in nearly pure mineral form) or mined in underground tunnels, collected in glory holes, and hauled by rail to the mill. A glory hole with its haulage tunnel is relatively small in comparison to an open pit, where the overburden (soil and rock without sufficient economic metal content) is removed to gain access to the ore-bearing strata. Today, most mines are either open pit, and/or have a network of underground tunnels, or both.

The first step before mining can begin is exploration, which is carried out today mainly by air, with sophisticated instrumentation, and covering large areas of the globe. Ground truthing, when the geology is promising, is carried out with borehole drilling and investigative trenching. Generally, these activities have low environmental impact. However, important environmental parameters could be collected in this early phase and later could be used to assess environmental consequences should a mine be developed. For example, water quality of boreholes and drainage characteristics of trenches could be used to plan for the future mine's waste rock pile and tailings basin siting. This potential is rarely utilized.

The second step is mining. It is the costliest of all activities, and one with a long-lasting environmental impact. It disturbs the hydrological conditions of the underground, contaminates the groundwater emerging from underground workings, and destroys surface landscapes with waste rock and tailings deposits.

The third step is mineral processing. This consists of first crushing the rocks to an even size and then grinding them in a ball mill. Once the desired sand-like size of the rocks is reached, physical separation methods, such as flotation, gravity separation, tabling, dense media, etc., are applied to the sand-water slurry. These methods result in a mineral concentrate that is further processed in the fourth step. The ground rock remaining after extraction of the target economic mineral is regarded as being of little or no economic value and is discharged, generally as a 30% sand-water slurry, known as tailings. The vast quantities of tailings and waste rock removed from the ground expose a very large surface area for weathering (oxidation) and hence represent long-term environmental liabilities, as evidenced by legacy sites worldwide.

The fourth step is collectively referred to as 'extractive metallurgical processes.' These processes differ from element to element, but take two forms, pyrometallurgy (smelting) and hydrometallurgy (leaching). The former has serious occupational health and safety issues in terms of gaseous emissions (air pollution) and slag (material left over after smelting). Slag is accumulated on land, but generally has not been considered a serious long-term environmental issue. Slags from smelting are extensively reused as building materials (Piatak et al., 2015). Most of the hydrometallurgical processes use chemical agents in relatively small quantities. The chemicals do not have nearly the long-term environmental consequences that tailings and waste rock piles have. In hydrometallurgical processes, only accidents during extraction are of concern. This is particularly true for the extraction of gold, which uses highly toxic cyanide. Although spills of tailings or process liquor are generally disastrous, the toxicity of cyanide is short-lived.

1.1 Weathering: Contaminant Generation – The Challenge

Minerals are the source of most of the elements present in all living organisms and are essential for growth and reproduction. Weathering, or the release of elements from rocks, occurs due to physicochemical forces (heat, wind, freezing, snow, rain, and erosion) and biogeochemical factors such as vegetation exudates and microbial activity (Gorbushina, 2007; Dontsova et al., 2020). These are primarily oxidative processes, driven by oxygen (air), water, and microbes. Microbes also bring changes in weathering/oxidation by accelerating the rate of oxidation by 1000 fold discussed in Chap. 8. These processes lead to the gradual breakdown of rocks and their minerals and supply elements to water and soil to support all life on the planet. Not all rocks display the same weatherability. The mineral composition of a rock and its weatherability are determined by a rock's history or genesis over geological timescales (also known as the rock cycle). The weatherability of rock determines the buffering capacity and elemental composition of the surrounding ground- and surface water, which, in turn, with climate, define the characteristics of ecozones within ecosystems (e.g., arctic or tropical) around the globe. Together with the growth and decay of vegetation, these processes govern the characteristics of surface water, groundwater and soil formation.

Exudates of higher plant roots alter the pH in the root zone and house microbes and fungi that assist in dissolving minerals in the soil to increase nutrient availability for plant growth (van Schöll et al., 2008). Lichens, fungi, and microbes grow attached to rock surfaces, exuding organic acids to liberate elements from minerals (Barker & Banfield, 1998; Uroz et al., 2009). These organisms control the availability of elements and determine the distribution of elements in water, air, and soil. An extensive discussion on weathering is given by Drever (2005) and Corenblit et al. (2011). Mine wastes represent a very large amount of exposed rock surface; much larger than the land area they occupy. Hence, weathering of the rock surfaces releases disproportionally greater amounts of elements to ground- and surface water, which for toxic metals can be detrimental to aquatic life.

Oxidation is primarily a geochemical reaction supported by microbes (Nordstrom, 2011). Many of these microbes belong to the group called Archaea (some of which are chemo-lithotrophs or rock eaters (Dave & Tipre, 2012; Singer & Stumm, 1970)), which derive energy from breaking mineral bonds, such as those in sulfidic minerals, increasing oxidation rates a thousand-fold (Dave & Tipre, 2012). Chemo-lithotrophic microbes are ubiquitous, colonizing any and all surfaces on the planet, and when they find the proper conditions, they flourish, as evidenced by the sulfate-oxidizing microbes in mine wastes. Further, the oxidation of sulfides in wastes generates heat (Blowes et al., 2003). This can lead to steaming/burning waste rock piles (Kuenzer & Stracher, 2012; Rosenblum et al., 2015). In the high arctic, the heat generated year-round prevents tailings from freezing completely (Elberling, 2004; Hollesen et al., 2011).

Exposed, excavated rocks in mine waste management areas are similar to natural, extreme environments dominated by sulfate and iron generated by volcanoes and their hot springs (King, 2003). In fact, life is hypothesized to have originated in these iron-rich environments (Deamer & Weber, 2010). Mine sites have many similar characteristics, and consequently the control of microbes should be the primary issue in mine waste management. Metal-bearing minerals are associated with the iron sulfides in the ore, waste rock, and tailings. When combined with atmospheric precipitation (i.e., carbonic acid) sulfuric acid, dissolves the minerals leaching metals from the rock. This is known as acid mine drainage (from tailings) or rock drainage (from waste rock). When the ore is contained in alkaline rocks, it is called neutral drainage.

These extreme ecosystems are controlled by weathering and ecological processes that are essentially the same the world over. Whether the wastes are in the desert, the tropics, on the Altiplano, or in the arctic, microbial, and fungal processes are the same the world over. Extremophile microbes have been found in the Atacama Desert, Altiplano, Patagonia, and in Antarctica and the Arctic (Orellana et al., 2018). Bio-geochemical cycles govern colonization, invasion, ecosystem evolution, and processes. Mine waste sites are no exception. They are merely environments where the native ecosystems have been scraped to bare rock. These sites are surrounded by a relatively undisturbed environment providing seed sources. Without any soils, these stark environments are controlled by oxidative processes. More natural ecosystems are a balance between oxidative and reductive processes.

The goal of the ecological engineer is to balance these overwhelmingly oxidative reactions with reductive processes. In pit lakes and drainage ditches this is accomplished by adding and/or abetting microbially-active, reductive sediments. In tailings, the area of weathering is the phreatic zone, the water saturation zone, in which the water level changes with the seasons. Here, the generation of weathering products has to be slowed. In waste rock piles, weathering products are produced throughout the pile, wherever oxygen penetrates with air into the rock voids. The weathering products are then transported out of the piles via atmospheric precipitation (Fig. 1.1). Since weathering occurs on the mineral surfaces, it is there that the oxidative processes must be slowed or stopped.

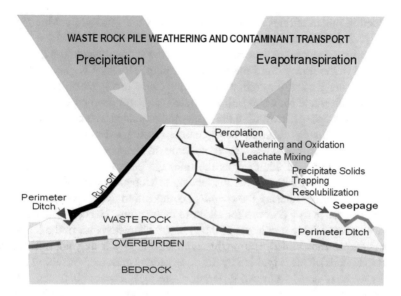

Fig. 1.1 A schematic cross section of water pathways within waste rock piles. Where the water percolates through the wastes, the metal acidity increases and precipitation and re-solubilization take place. Water creates its own selective pathways through the pile. It follows that to precipitate the weathering products *in-situ*, the reactant needs to be carried by rain along selective pathways

1.2   Difficulties in Predicting Contaminant Generation

Once an ore body is evaluated as viable, possible effluents are characterized using a procedure called Acid-Base Accounting (ABA). However, predicting drainage or drainage water characteristics is difficult. Nevertheless, when an ore body is confirmed, and a mining operation is planned, government regulators request forecasts of the acid generation potential of the wastes generated. Many test procedures have been developed over time to improve accuracy. A detailed review of ABA techniques is presented by Dold (2017), with an emphasis on mineralogy. Drainage from a waste rock pile generally emerges in an oxidized form, whereas tailings drainages emerge reduced, with low E_h and circum-neutral pH. When exposed to oxygen in the seeps or drainages, metal contaminants rapidly precipitate. This reaction produces hydrogen ions which decrease the pH (details in Dold, 2014).

Generally, ABA test work is based on ground- or segregated- rock. Segregation (based on particle size and mineralogy) separates reactive waste from non-reactive rock. This classification is supposed to ensure a relatively homogenous distribution of all rock types occurring in the mine wastes, with both neutralizing and acid-generating minerals. Grinding the segregated rocks creates a relatively homogeneous sample for the ABA tests, as well as allowing more exposure to the mineral surfaces. While neutralizing and acid-generating minerals may be homogeneously distributed in milled tailings for the ABA tests, they are generally not homogenous in waste rock and the tailings.

Weathering products are generated mainly in the vadose zone of the fine-grained tailings and in the flow path of the waste rock piles. Drainage is generated only at locations in the wastes through which atmospheric precipitation passes. In the stockpiles, both waste rock and tailings, water can form perched water tables and develop distinct flow paths. Within the paths, contaminated water encounters different minerals, causing different precipitation reactions. Neutralizing and acid-generating rocks do not release their minerals at the same rate, which reduces possible interactions. In addition, as the drainage passes through the stockpiles, internal chemical precipitation occurs, leading to secondary minerals, some highly water-soluble, further altering the chemical composition of the emerging drainage. ABA procedures hardly account for all these interactions. Hence, ABA test procedures are rarely reliable predictors of the drainage characteristics in the long-term.

For example, rock mineralization throughout the stockpile may not reflect that of the tested material. This is typified by the observations presented in Table 1.1. Water samples were collected from two different waste rock seepages between 1992 and 1994 and analyzed by Inductively Coupled Plasma spectroscopy (ICP) for their

Table 1.1 Water quality differences in seepages from the northwest and southeast slopes of a waste rock pile

	Northwest Toe Seepages					Southeast Toe Seepages				
	Avg.	S.D.	Min	Max	n	Avg.	S.D.	Min	Max	n
Temp (°C)	9.3	5.3	0.8	23	54	9	5.9	0.7	21	66
pH	4.1	3.8	3	6	58	3.2	2.7	1.9	6.3	69
Cond (μS cm^{-1})	1410	644	550	4620	58	1382	874	273	4550	69
Acidity (mg.L^{-1})	116	156	4.7	653	32	177	353	10	1723	35
TDS	1427	583	362	2490	35	1372	717	327	4290	42
TSS	25	47	<1	110	5	425	790	<1	2300	9
Al	0.77	0.66	0.11	2.2	10	16	27	0.30	81	15
Diss As	81	102	0.01	520	47	32	30	0.06	130	63
Tot As	73	61	0.07	230	28	27	22	0.18	69	37
Ba	0.03				1	0.01	0.01	0.01	0.02	3
Ca	159	58	54	290	38	176	55	66	292	47
Diss Fe	0.12	0.23	0.001	1.0	32	16	51	0.001	220	42
Tot Fe	0.60	0.62	0.05	2.0	11	34	70	0.02	230	11
K	30	7.3	11	45	38	28	9.8	12	56	47
Mg	74	29	20	147	38	87	49	31	253	46
Mn	5.9	3.9	2.6	17	22	8.4	3.8	4.2	19	32
Na	31	11	13	70	38	23	6.7	11	35	47
Diss Ni	31	11	13	70	38	23	6.7	11	35	47
Tot Ni	88	75	2.6	400	47	85	94	1.8	470	62

The waste rock was declared as non-acid-generating based on the standard tests elemental composition. Large variations in drainage composition were found between different slopes of a waste rock pile. One slope (SE) was exposed toward the sun and the other more shade (NW). The waste rock was tested using a series of acid-base-accounting procedures and declared non-acid producing waste rock. Yet, both slopes produced acidic drainage ranging from pH 2 to 6 (SE) and 3–6 (NW), respectively (Table 1.1). The observations in Table 1.1 support the contention that prediction of drainage quality is fraught with difficulties. It follows that more reliable approaches should be developed and considered.

These forecasts might be made more accurate by constructing and monitoring rock piles during exploration. For example, ore and waste rock exposed during exploration could be left as test areas to be monitored over time. If drill holes collect water they could be used to monitor its quality and after they are pumped dry checked occasional for water quality or its dry state over the life time of the mine. This would provide useful information at close out of the mine waste and water management area. Between exploration and development of a mine, years often pass. The monitoring of these weathered exploration cavities, trenches or drill core racks might give a more accurate characterization of the drainage. Finally, when operations start, large, outdoor test piles may provide additional, useful data for waste management during operations and planning for decommission.

References

Barker, W. W., & Banfield, J. F. (1998). Zones of chemical and physical interaction at interfaces between microbial communities and minerals: A model. *Geomicrobiology Journal, 15*(3), 223–244. https://doi.org/10.1080/01490459809378078

Blowes, D. W., Ptacek, C. J., Jambor, J. L., & Weisener, C. G. (2003). The geochemistry of acid mine drainage. In B. S. Lollar (Ed.), *Treatise on geochemistry* (Vol. 9, pp. 149–204). Elsevier-Permamon.

Corenblit, D., Baas, A. C., Bornette, G., Darrozes, J., Delmotte, S., Francis, R. A., Gurnell, A. M., Frédéric, J., Naiman, R. J., & Steiger, J. (2011). Feedbacks between geomorphology and biota controlling Earth surface processes and landforms: A review of foundation concepts and current understandings. *Earth-Science Reviews, 106*(3–4), 307–331.

Dave, S. R., & Tipre, D. R. (2012). Coal mine drainage pollution and its remediation. In T. Satyanarayana, B. N. Johri, & A. Prakash (Eds.), *Microorganisms in environmental management* (pp. 719–743). Springer.

Davis, R. A., Welty, A. T., Borrego, J., Morales, J. A., Pendon, J. G., & Ryan, J. G. (2000). Rio Tinto Estuary (Spain): 5000 years of pollution. *Environmental Geology, 39*, 1107–1116.

Deamer, D., & Weber, A. L. (2010). Bioenergetics and life's origins. *Cold Springs Harbor Perspectives in Biology, 2*(2), a004929.

Dold, B. (2014). Evolution of acid mine drainage formation in sulfidic mine tailings. *Minerals, 4*, 621–641. https://doi.org/10.3390/min4030621

Dold, B. (2017). Acid rock drainage prediction: A critical review. *Journal of Geochemical Exploration, 172*, 120–132.

Dontsova, K., Balogh-Brunstad, Z., & Chorover, J. (2020). Plants as drivers of rock weathering. In *Biogeochemical cycles* (pp. 33–58). https://doi.org/10.1002/9781119413332.ch2

Drever, J. I. (Ed.). (2005). *Surface and ground water, weathering, and soils: Treatise on geochemistry* (Vol. 5). Elsevier. Available online.

Elberling, B. (2004). Disposal of mine tailings in continuous permafrost areas: Environmental aspects and future control strategies. In *Cryosols* (pp. 677–698). Springer. Available online.

Gorbushina, A. A. (2007). Life on the rocks. *Environmental Microbiology, 9*(7), 1613–1631. https://doi.org/10.1111/j.1462-2920.2007.01301.x

Hollesen, J., Elberling, B., & Jansson, P. E. (2011). Modelling temperature-dependent heat production over decades in High Arctic coal waste rock piles. *Cold Regions Science and Technology, 65*(2), 258–268.

King, G. M. (2003). Contributions of atmospheric CO and hydrogen uptake to microbial dynamics on recent Hawaiian volcanic deposits. *Applied and Environmental Microbiology, 69*(7), 4067–4075.

Kuenzer, C., & Stracher, G. B. (2012). Geomorphology of coal seam fires. *Geomorphology, 138*(1), 209–222.

Nordstrom, D. K. (2011). Hydrogeochemical processes governing the origin, transport and fate of major and trace elements from mine wastes and mineralized rock to surface waters. *Applied Geochemistry, 26*(11), 1777–1791.

Piatak, N. M., Parsons, M. B., & Seal, R. R. (2015). Characteristics and environmental aspects of slag: A review. *Applied Geochemistry, 57*, 236–266.

Ritter, S. K. (2014). Making the most of red mud. *Chemical Engineering News, 92*(8), 33–35. Retrieved from http://cen.acs.org/articles/92/i8/Making-Red-Mud.html

Orellana, R., Macaya, C., Bravo, G., Dorochesi, F., Cumsille, A., Valencia, R., Rojas, C., & Seeger, M. (2018). Living at the frontiers of life: Extremophiles in Chile and their potential for bioremediation. In *Frontiers in Microbiology* (Vol. 9, p. 2309). https://www.frontiersin.org/article/10.3389/fmicb.2018.02309

Rosenblum, F., Finch, J. A., Waters, K. E., & Nesset, J. E. (2015). A test apparatus for studying the effects of weathering on self-heating of sulfides. In *COM 2015, The Conference of Metallurgists, Canadian Institute of Mining, Metallurgy and Petroleum.*

Singer, P. C., & Stumm, W. (1970). Acidic mine drainage: The rate-determining step. *Science, 167*(3921), 1121–1123. https://doi.org/10.1126/science.167.3921.1121

Strosnider, W. H. J., López, F. L., & Nairn, R. W. (2011). Acid mine drainage at Cerro Rico de Potosí I: Unabated high-strength discharges reflect a five-century legacy of mining. *Environmental Earth Sciences, 64*(4), 899–910.

Uroz, S., Calvaruso, C., Turpault, M. P., & Frey-Klett, P. (2009). Mineral weathering by bacteria: Ecology, actors and mechanisms. *Trends in Microbiology, 17*(8), 378–387.

van Schöll, L., Kuyper, T. W., Smits, M. M., Landeweert, R., Hoffland, E., & Van Breemen, N. (2008). Rock-eating mycorrhizas: Their role in plant nutrition and biogeochemical cycles. *Plant and Soil, 303*(1–2), 35–47.

Wise Uranium Project. (2019). https://www.wise-uranium.org/mdafbr.html

Chapter 2
Dimensions of Global Mining Waste Generation and Water Use

Michael P. Sudbury

Abstract Mining and mineral processing are essential to our industrialized world. However, the global dimensions of waste generation by mining activities are difficult to assess. Based on global statistics, the chapter not only illustrates the extent of the challenges, but also projects that land usage for mining and agriculture appear to be on a collision course. Given the scarcity of reliable numbers and the uncertainty of the underlying assumptions, the chapter cannot give exact figures, but provides an educated estimate of land and water resources consumed by mining and its wastes, globally. These estimates suggest that changes in mine waste and water management approaches are needed now and by future generations. A paradigm shift in land use, water usage, and mining techniques will not only benefit society, but is essential for the continued extraction of metals, rare earth elements, and other mineral resources.

Keywords Conflict of interest · Arable land use · Water resources · Desalination

Mining and mineral processing are vital activities in an industrialized world, but their activities are mostly conducted in locations relatively remote from urban society, thereby attracting little attention except when, on relatively rare occasions, a major incident, such as a tailings dam failure or a rock failure, attracts wide media attention. The industry has developed procedures to minimize the risk of such incidents. A less spectacular but increasingly important aspect of mining is the need for land to store wastes, and the need for water to transport and process ores, especially when these needs compete with a limited area of arable land and/or water supplies for irrigation. This competition is becoming more intense as the world's population increases, requiring more mineral resources and water, and requiring more agricultural production from a finite arable land area.

M. P. Sudbury (✉)
Michael P. Sudbury Consulting Services Inc., Oakville, ON, Canada
e-mail: msudbury@cogeco.ca

© The Author(s), under exclusive license to Springer Nature
Switzerland AG 2022
M. Kalin-Seidenfaden, W. N. Wheeler (eds.), *Mine Wastes and Water,*
Ecological Engineering and Metals Extraction,
https://doi.org/10.1007/978-3-030-84651-0_2

This section is a first attempt to put together a set of global statistics to quantify, at least approximately, the dimensions of this competition and to identify some of the ways that mining, and agriculture might cooperate to the mutual benefit of society in general. The quantification of these dimensions is complex, but approximations are adequate to define the global challenge and to put mining activities into a global perspective. Worldwide estimates are offered, based on the scant data available in the literature. Where no literature has been found, estimates are made using the experience gained by the author over a lifetime in key positions in mining companies worldwide. What's important here is not an exact figure, but an educated estimate of the order of magnitude of tonnages of waste rock and tailings produced with the concurrent use of freshwater. The global land use is presented as waste generation in units that are relatively easily to comprehend.

2.1 Dimensions of Global Mine Waste Generation and Water Consumption

There are many organizations worldwide that collect statistics on mining wastes, including the UN Statistics Division (2021) and the U.S. Geological Survey (2021), which provide global production figures for minerals and metals. Additionally, mining companies often provide information on solid waste production and water use in annual environmental or sustainable development reports. The data are framed within the global mining context covering a 25-year period, as this is a typical mine lifespan. Emphasis is placed on water resource supply and use by mining operations that appear to create competition, as they represent a close link to land use for agriculture. Abandoned or orphaned mine sites and their wastes, mankind's shared global historic inheritance, are not considered in these estimates.

Population growth, increased prosperity, resource demand, and resource competition are forcing the mining industry to rethink the future of mineral extraction. These factors are inextricably linked, with consequences for people and the future. Which should be emphasized—irrigation for agriculture, or mine water and waste (Bebbington & Williams, 2008)?

The population of the globe is predicted to increase from the current 7.8 billion (2021) to 9.2 billion by 2040—an increase of about 18% (Worldometer, 2021). This will increase the demand for food, water, and raw materials of all kinds, including minerals and metals. Global freshwater consumption per capita increases in proportion to the increase in per capita income (UN Water, 2021a). The global average per capita income has been increasing exponentially since the start of the Industrial Revolution and is expected to increase by 33% in the next 25 years (The Maddison-Project, 2013). This will result in a 60% increase in global water consumption. This estimate discounts major natural or human-generated disasters.

Water is vital in many mining operations, as it is used for dust control, drilling, transportation of solids, furnace cooling, and quenching slag and off gassing (H_2S

emissions from the tower of the refinery) as well as in refining operations (Mudd, 2008). Efforts are underway to reduce freshwater use in mining, but it is not yet common practice (Bruce & Seaman, 2014).

2.2 Global Mine Water Usage – Annual Estimates

With 80 km³/y (Table 2.1), the industry's global water consumption is relatively small compared to other industrial sectors (Table 2.2; UN Water, 2021a, b; Ecological Society of America, 2001). However, this consumption leads inevitably to contamination of groundwater and surface water. Increased surface area of waste rock and tailings exposed to weathering releases not only soluble elements, but also large quantities of suspended solids. Furthermore, during mine development, dewatering is often needed to access the ore to be mined, which in turn may deplete freshwater aquifers. The current global use of water of all major industries is estimated to be around 4500 km³ per year (Table 2.2), or about 10% of the net precipitation (rain and snow) falling on land.

Currently, the water supplied by atmospheric precipitation is supplemented by draw-down of aquifers. Depletion of aquifers will increase competition for surface water supplies and lead to increased reliance on desalination plants in arid locations. These estimates highlight the fact that mining and milling might well be on a

Table 2.1 Global water use estimates for the commodities listed, not including REE

Mineral	Coal	Ferro-alloys	Iron ore	Gold	Copper	Oil sands	Other	Total
Ore (10^6t.)	7000	1000	2600	1700	1100	1000	5700	20,100
Process Water ($m^3.t^{-1}$)	2	1	5	5	5	10	5	
Water Use (10^9 t.y^{-1})	14	1	13	8.5	5.5	10	28.5	80.2

Source: UN Water (2021a, b)

Table 2.2 Water consumption estimates of global industrial sectors

Industry	Irrigation	Forest	Power	Desalination	Municipal	Mining	Manufacturing	Global Use	U.S.A.
Product	Food	Paper	Heat sink	Freshwater	Services	Metals	Various		
Water Use (10^9t.y^{-1})	1500	10	70	(−28)	500	80	2340	4500	*2000*
Per Person ($m^3.d^{-1}$)	0.6	0.004	0.03	(−0.01)	0.2	0.03	0.9	1.76	*5.5*

Source: UN Water (2021a, b)

Table 2.3 Estimates of global mined tonnages of minerals and their associated wastes

Material	Coal Ferro-alloys [1]	Ferro-alloys*	Iron ore	Gold	Base metals	Oil sands	Industrial-agricultural minerals	Total World
Ore (10^6 t/y)	7000	1000	2600	1700	1200	1000	5700	20,100
Waste rock/ tailings (10^6 t/y)	7000	1000	1000	3000	2100	1000	5700	20,800

Note: *Ferro-alloy ores include nickel laterites and beach sands

collision course with civilization's water requirements, given the well-documented global water scarcity (UN Water, 2021b). Concomitantly, the dollar value of water will increase and the degradation of resources will gain importance. In Table 2.3, the global mined tonnages of minerals and their associated wastes are estimated. The global area covered annually with mine wastes can be estimated to be on the order of 1000 km^2, assuming an average loading of 20 tonnes.m^{-2}.

2.3 Global Mine Lands Usage Annual Estimates

Global satellite imagery might provide a more substantive and definitive estimate. The estimated annual total land area committed to mining is small compared to the total land area of the Earth, which is one hundred and forty-nine million square kilometers (World Atlas, 2021). However, it is large enough to be a serious local issue and becomes even larger if extrapolated over a century or more. A time trend of the available arable land area with the growing world population is presented in Fig. 2.1.

This emphasizes that arable land is being lost at an unsustainable pace. For example, between 1950 (0.52 ha/capita) and 2050, it is estimated that 0.36 ha per person will be lost (UN FAO, 2009). It can be expected that conflicts between agriculture and mines and their wastes will increase (Hilson, 2002). Already there are localized conflicts, as some groups like Mining Watch (Mining Watch Canada, 2021) document the conflicts between Canadian mining companies and local landscapes. Mine wastes not only consume land but also create dust storms and silt streams, and contaminate surface water and/or groundwater. Failure of tailings dams is the cause of many disasters, as the long-term stability of dams is an acknowledged engineering challenge (WISE Uranium Project, 2021).

The rising global population is placing increased pressure on the finite area of arable land for food production and will increase the demand for irrigation (Table 2.4). The area of arable land is also shrinking as an increasing population requires more land for infrastructure. Some forest lands could be converted to arable land, but this would bring about a loss of carbon dioxide sinks, a loss of water-holding capacity, loss of wildlife habitat, and increased erosion, producing

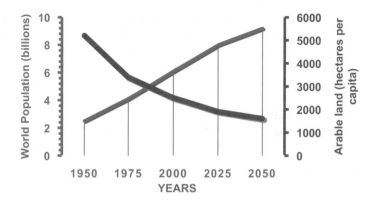

Fig. 2.1 Time trend of available arable land area with population growth
Source: Fast facts: The state of the world's land and water resources. (FAO, 2011).

Table 2.4 Cost of desalination plants

Year	1982	1992	2002	2012	Base cost*
Cost (US$.m^{-3})	1.5	1.1	0.7	0.6	0.82
Energy (kwh.m^{-3})	8.1	5.3	4.5	3.8	2.2–3.0

Note: *The cost of desalinating sea water has been falling and desalination plants are being constructed by mining operations to avoid competition with scarce local water supplies
Source: Soruco and Philippe (2012)

desertification and loss of livelihood for aboriginal peoples. This alternative is not generally regarded as desirable or viable. Some grassland may be suitable for crops, but usually only with irrigation.

Many mines exist in the tropical or subtropical deserts that cover a total area of 15.3 million km^2. The combined area of tropical and subtropical deserts is the same as the global area of arable land, and with an adequate water supply, could presumably be equally productive (Wikipedia, 2021). In these arid environments, water supply and effluent discharge are major issues that must be balanced against lucrative ore bodies that are mined in the same areas. During mine operation, the local community can benefit from sharing a supply of freshwater, as the mining industry is gradually adopting desalination to guarantee a supply of freshwater for ore processing.

The base metal mining industry is also an intensive user of water, commonly requiring about 3 tonnes of water per tonne ore. More intensive processing to extract and upgrade more minerals will likely increase this quantity, but not by a significant amount. Control, recycling, and purification of process water and tailings/waste rock run-off and seepage will, however, become increasingly important, and ecological engineering has an important role to play in this endeavor.

Currently, copper mines in arid regions, notably in the Andes, are resorting to desalination of Pacific Ocean water for mine water supplies at a cost, including pumping, of US$6–8 per tonne. Countries with sub-tropical climates (high solar energy at ground level) including India, China, Saudi Arabia, Australia, Morocco, Namibia and the SW USA, are all working to develop economically viable concentrated solar power systems. An 'all in' cost of under 10 cents per kWh is forecast by 2030 and with concurrent improvements in reverse osmosis systems (feed water purification & graphene diaphragms) production costs in the US of $0.50–1.00 range are predicted (IRENA, 2016).

World Metal Institute statistics indicate the value of non-ferrous metal production in developed (water-rich) countries are over three times greater per square km than in desert countries. While open to many explanations, lack of water is an important factor (Reichl et al., 2017).

The total cost for a large desalination plant ($20,000 \text{ m}^3.\text{d}^{-1}$), with power costs at 10 cents per kilowatt-hour, can be approximated as $1 per cubic meter (Table 2.4). This cost applies to the plant capital and operating costs only, and does not include the cost of delivering seawater, returning brine, and delivering desalinated water to the point of use. It should be noted that pipeline capital and operating costs to deliver freshwater to a mine at a high elevation can triple the final delivered cost (Soruco & Philippe, 2012).

References

Bebbington, A., & Williams, M. (2008). Water and mining conflicts in Peru. *Mountain Research and Development, 28*(3), 190–195. Retrieved from http://snobear.colorado.edu/Markw/Research/08_peru.pdf

Bruce, R., & Seaman, T. (2014). *Reducing freshwater use in the production of metals.* Teck Resources Limited. Retrieved from https://www.teck.com/media/CESL-Publication-Copper-reducing-fresh-water-use-in-the-production-of-metals.pdf

Ecological Society of America. (2001). Water in a changing world. *Issues in Ecology #9.* http://www.esa.org/esa/wp-content/uploads/2013/03/issue9.pdf

Hilson, G. (2002). An overview of land use conflicts in mining communities. *Land Use Policy, 19*(1), 65–73. https://www.sciencedirect.com/science/article/abs/pii/S0264837701000436?via%3Dihub

IRENA. (2016). The power to change: Solar and wind cost reduction potential to 2025, ISBN 978-92-95111-97-4. 8. https://www.irena.org/publications/2016/Jun/The-Power-to-Change-Solar-and-Wind-Cost-Reduction-Potential-to-2025

Mining Watch Canada (2021). Agriculture and mining land conflicts. www.miningwatch.ca

Mudd, G. M. (2008). Sustainability reporting and water resources: A preliminary assessment of embodied water and sustainable mining. *Mine Water and the Environment, 27*(3), 136–144.

Reichl, C., Schatz, M., & Zsak, G. (2017). *World mining data* (Minerals production) (Vol. 32). International Organising Committee for the World Mining Congresses.

Soruco, L., & Philippe, R. (2012). Upcoming trends in water supply costs for copper mining in arid regions. In Fernando Valenzuela & Jacques Wiertz (Eds.), Water in mining 2012: Proceedings of the 3rd international congress on water management in the mining industry, Santiago, Chile.

The Maddison Project. (2013). Global per capita income. https://www.rug.nl/ggdc/historicaldevelopment/maddison/releases/maddison-project-database-2013

United Nations, FAO. (2009). *Global agriculture towards,* 2050. http://www.fao.org/fileadmin/templates/wsfs/docs/Issues_papers/HLEF2050_Global_Agriculture.pdf

United Nations, FAO. (2011). *The state of the world's land and water resources for food and agriculture (SOLAW) – Managing systems at risk. Food and Agriculture Organization of the United Nations*. Rome and Earthscan.
United Nations, Statistics Division. (2021). Mining statistics. https://unstats.un.org/unsd/envstats/qindicators.cshtml
United Nations, UN Water. (2021a). Water usage facts. https://www.unwater.org/
United Nations, UN Water. (2021b). Water scarcity. https://www.unwater.org/water-facts/scarcity/
US Geological Survey. (2021). Mining statistics. https://www.usgs.gov/centers/nmic/publications
Wikipedia. (2021). Desert farming. https://en.wikipedia.org/wiki/Desert_farming
WISE Uranium Project. (2021). Tailings dam failures. http://www.wise-uranium.org/help.html
World Atlas. (2021). Total land area of Earth. https://www.worldatlas.com/geography/planet-earth.html
Worldometer. (2021). World population. https://www.worldometers.info/world-population/

Chapter 3
Toward a Sustainable Metals Extraction Technology

Bryn Harris (iD)

Abstract Since metals are essential in modern society, cost-effective, sustainable remediation measures need to be developed. Engineered covers and dams enclose wastes and slow the weathering process, but, with time, become permeable. Neutralization of acid mine drainage produces metal-laden sludges that, in time, release the metals again. These measures are stopgaps at best, and are not sustainable. Focus should be on inhibiting or reducing the weathering rate, recycling, and curtailing water usage. The extraction of only the principal economic mineral or metal generally drives the economics, with scant attention being paid to other potential commodities contained in the deposit. Technology exists for recovering more valuable products and enhancing the project economics, resulting in a reduction of wastes and water consumption of up to 80% compared to "conventional processing." Implementation of such improvements requires a drastic change, a paradigm shift, in the way that the industry approaches metals extraction. Combining new extraction approaches, more efficient water usage, and ecological engineering methods to deal with wastes will increase the sustainability of the industry and reduce the pressure on water and land resources.

Keywords Chloride metal extraction · Sustainability mining wasterock ·
Endangered elements · Intrinsic energy of sulfides

The metals extraction industry is now facing possibly its greatest-ever challenge, with the need to demonstrate "sustainability" in the face of dwindling reserves and grades, increased restrictive legislation, and increasing costs. To even entertain the idea of being "sustainable" in the face of being essentially nonrenewable, the industry theoretically can no longer afford to throw away up to 99% of the material it mines, the act of mining being the largest single cost of getting the mineral-bearing

B. Harris (✉)
Alexandria, ON, Canada
e-mail: bryn@sutekh.org

© The Author(s), under exclusive license to Springer Nature   17
Switzerland AG 2022
M. Kalin-Seidenfaden, W. N. Wheeler (eds.), *Mine Wastes and Water,*
Ecological Engineering and Metals Extraction,
https://doi.org/10.1007/978-3-030-84651-0_3

rocks. Mining, as opposed to processing (grinding and concentrating), represents generally the main cost associated with any metals extraction project. There are some polymetallic ore bodies, but at most, only one or two metals are extracted, dictated by the market value of the metal.

Of the various extraction technologies, a brief mention should be made of bacterial leaching for base metals, as it is often perceived as being the "environmental solution." This is because it uses "natural processes," but it is in many ways worse than conventional approaches. While bacterial leaching has seen some measure of success in uranium and gold plants, attempts to apply the technology to base metals have been largely unsuccessful. An initial pilot project in Chile, comprised of a joint venture between BHP Billiton and Codelco, known as Alliance Copper, ultimately resulted in the building of a 20,000 tpa (tonnes per annum) copper plant (Batty & Rorke, 2006). However, the project was terminated in October 2006, having not achieved its objectives. Similar processes were tried for nickel and zinc, with similar results, and the failures likely had similar causes, given that an understanding of microbial processes was lacking or was not utilized. Microbial systems are considered difficult to control.

However, more recently, there have been two projects which attained commercial operation, albeit briefly in one case, for nickel. Of these, Mondo Minerals Nickel in Finland attained commercial operation on a waste stream from a talc operation (Neale, et al., 2016a, b), but is now on care and maintenance, and Terrafame is reactivating a project in Finland (Heikkinen & Korte, 2019). Terrafame is the only primary process for processing nickel arisings, although it rides very strongly on the back of a vibrant conventional zinc operation.

In general terms of metal extraction and recovery, the industry has a well-deserved image of being slow and conservative, although it is not totally averse to change and innovation. There have been some exciting and innovative processes developed since the end of World War II, notably:

- Pressure leaching (leaching at higher than atmospheric pressures and higher temperatures than boiling water)
- Ammonia-based processes for Ni, Co, and Cu pioneered by the Canadian company Sherritt Gordon (now known simply as Sherritt or Sherritt International)
- Oxidation of zinc sulfide concentrates in sulfuric acid, also pioneered by Sherritt
- Oxidation of refractory gold concentrates in sulfuric acid to make the gold amenable to subsequent cyanidation leaching, pioneered by Barrick and others
- Nickel laterites in sulfuric acid (although this has not yet been proven as being generally economically-viable)
- Copper solvent extraction in various forms using "designer" complex organic carbon molecules
- CIP/CIL (carbon-in-pulp or carbon-in-leach) for gold and silver recovery, using activated (usually coconut shell) carbon to preferentially absorb the gold or silver

- Falconbridge (now Glencore) chlorine leach process (Falconbridge is one of a few plants that make use of chloride chemistry).

These, while in themselves highly commendable and successful, have unfortunately not addressed the basic issues confronting the industry today, namely sustainability and environmental liability. It is considered, therefore, that the industry needs to completely change its mindset and how it operates if it is to remain both competitive and at the same time to reduce or possibly eliminate environmental liability, and be "sustainable." Given the actual costs of mining itself (getting to the ore body, breaking the rock and hauling it to the surface), and that large, rich ore bodies are no longer being found, then it surely makes both economic and sustainable sense to maximize the recovery of all metals that have value and have been mined. The question, therefore, is obvious: why is the industry not extracting more out of the mined rocks?

Throughout the 1970s and 1980s there was a great deal of almost evangelical interest in and enthusiasm for, and research into, developing hydrometallurgical processes for the treatment of (primarily) copper sulfide concentrates (Paynter, 1973; McLean, 1982; Flett et al., 1983; Wadsworth, 1984). An astonishing number of different processes emerged during this time, with so-called "out of the box" thinking. Great emphasis was placed on chloride-based processes, attempting to take advantage of the many unique properties of chloride chemistry, although several sulfate-based circuits were also conceived, together with one notable ammonia-based plant, Anaconda's Arbiter Plant (Kuhn et al., 1974). Unfortunately, only two of the processes achieved commercial operation (Duval's chloride-based CLEAR – Copper Leach, Electrowinning and Regeneration - and Arbiter), and then only for a limited time, thus leading to a general suspicion of "new technologies" that continues to exist.

In 2003, at a major international hydrometallurgy symposium, the keynote paper addressed why new hydrometallurgical processes failed (Halbe, 2003). Four important aspects were highlighted, namely, (i) if any pilot-scale testing was conducted, it was to generate product, not to confirm process parameters; (ii) equipment was downsized or design criteria were made less conservative in response to projected cost overruns; (iii) process flowsheets were unusually complex, with prototype equipment in two or more critical unit operations; and (iv), somewhat surprisingly, there was a lack of understanding of the process chemistry. Consequently, very few of the processes even reached the pilot stage, with the unfortunate result that there is now an inherent distrust of any new extraction processes or technology. This distrust has only been enhanced by the failure of four HPAL (high-pressure acid leach) projects in Western Australia since the mid-1990s, and more recently, the failed hydrochloric acid regeneration plant of SMS Siemag (now SMS Group) at the (formerly) Thyssen Krupp steel plant in Calvert, Alabama.

3.1 Estimating the Full Extraction Potential of Mined Rock

Referring to the fact that mining costs represent a large and significant part of any overall metals or industrial minerals project cost, a hypothetical example is given below where it would make both economic and environmental sense to maximize the recovery of all metals that have value and have been mined, a desirable step toward sustainability of the mining industry. Consider that a nickel laterite, with a composition of 1.2% Ni, 0.1% Co, 5% Al, 15% Mg, and 30% Fe is processed with 90% recovery of Ni, Co, and Fe, and 75% recovery of Al and Mg. Taking prices (in US dollars, 2018) of $5/lb for Ni, $10/lb for Co, $0.2/lb for Al_2O_3, $40/tonne for Fe_2O_3, and $50/tonne for MgO, the following revenues are generated for a plant nominally producing 50,000 tonnes of LME (London Metal Exchange) grade Ni:

- Ni—$550 million
- Co—$90 million
- Al_2O_3—$150 million (350,000 ton)
- Fe_2O_3—$70 million (1.8 million ton)
- MgO—$30 million (610,000 ton)

By this analysis, the revenues of the project could be increased significantly over those generated simply by nickel (and cobalt). Furthermore, there are additional benefits in that there are close to 2 million tonnes of residues (equivalent to approximately 50% of the material originally mined) that will not have to be disposed of, and hence an appreciable reduction in mining wastes. There are also indirect savings and benefits, particularly from an environmental viewpoint, in that water is saved because it is not used for mining the equivalent tonnage of Al, Fe, or Mg from a primary ore body, such as bauxite, iron ore, or magnesite/dolomite mines, and the tailings that would necessarily be produced from such mining would no longer be generated. Furthermore, there is always a premium for high-grade hematite, which does not need further processing, which also occurs in nickel laterite deposits, so that the revenues to be derived for this hypothetical mine are probably significantly understated.

However, in the context of the illustration, the actual prices are irrelevant, since the objective is simply to demonstrate the points that these values have been mined, but, with traditional processing methods and especially mindsets, they are not only not being realized, but also are being disposed of, thereby creating an environmental problem that can, and should, be avoided (Dry, 2015).

3.2 Barriers to Higher Recovery of Metals from Mined Rock

The industry, because of the past failures noted above, and being generally reluctant to embrace any sort of risk or major change, has standard arguments formulated against such an approach for all or more metal extraction.

1. The technology to achieve the recoveries in sufficiently pure form does not exist.
2. If it did exist, then it would be too expensive and difficult to implement, especially as a retrofit, that is, into the existing process equipment.
3. The existing markets could not absorb additional tonnages.

For the case of iron and aluminum, the amounts generated from the hypothetical laterite project are sufficient to operate a stand-alone steel mini-mill and aluminum smelter. This ought to be attractive in an established and diverse mining area, such as Western Australia. There is clearly sufficient aluminum associated with the Western Australian laterites to sustain the existing local aluminum smelters, resulting in less bauxite needing to be mined and imported.

For environmental technologies, that is, technologies that deal with existing tailings ponds, and especially those based on ecological engineering principles, points 1 and 2 above apply. It should also be pointed out in reference to point 3 that even if 100% of the iron associated with current non-ferrous metal mining was recovered, it would still be <10% of global iron production.

3.3 Future Resources: Old Legacies, the Ocean and the Sky

To put the above into context, our resource-hungry world needs to realize that commodities are under serious threat. Table 3.1 (abstracted from data of the 2019 edition of the US Geological Survey Mineral Commodities (USGS, 2019) shows that several common metals, whose availability we take for granted, that with known, identified reserves, and at current (i.e., no increase in) consumption rates, then there are less than forty years of supply left. Demand for all of these metals will increase, however, especially for cobalt, which is a key component in electric vehicles (EVs), the growth of which will be substantial in the next decade, so that the estimated number of years will, in actuality, be somewhat lower. Whilst we can expect some new reserves to be identified, it is clear that in the 21st century, humankind will face a crisis in the supply of the very metals that underpin our society. This table indicates only some of the commodities under threat, however, The Royal Society of Chemistry has generated an innovative, colour-coded Periodic Table highlighting the elements at risk, as shown in the Fig. 3.1 (Royal Society of Chemistry, 2011), albeit less up to date. Sackett has elaborated on these risks, and what they mean for the human race, in some detail (Sackett, 2012).

Table 3.1 Metal supplies with identified reserves

Metal	Cobalt	Copper	Gold	Lead	Nickel	Silver	Zinc
Production (2018) tonnes	140,000	21 million	3,260	4,400,000	2,300,000	27,000	13 million
Identified Reserves Beyond 2019 tonnes	6,900,000	830 million	54,000	83 million	89 million	560,000	230 million
Years Remaining Beyond 2019	49	39	17	19	39	21	18

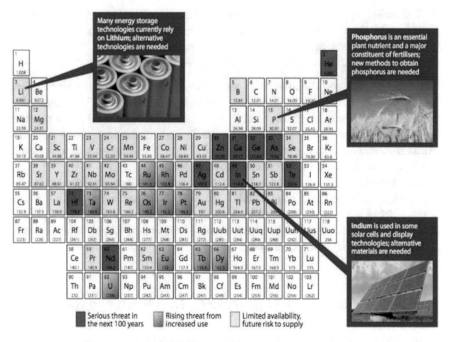

Fig. 3.1 Periodic chart of the elements at risk. Royal Chemistry [1] and the chemistry Innovation Knowledge Transfer Sustainable Network's Technology Roadmap

This projected shortage is, however, an opportunity from an environmental perspective, and especially with old tailings dams. Mining and extraction were not very efficient in the past, even a century ago, and these old dumps, which are already an environmental liability, contain billions of dollars of metals that are recoverable with more modern extractive processes. One such already-mined resource in Canada is the Sudbury Tailings from the operations of both Vale (Inco) and Glencore (Falconbridge) (Dry & Harris, 2010; Peek et al., 2011; Harris & Dry, 2020). Since the early 1990s alone, Glencore has disposed of 8 million tonnes (dry basis), with an average grade of ~0.8% Ni (Peek et al., 2011). Cobalt content was not given, but unpublished data have shown 0.03-0.05% Co and nickel contents as high as 1.2%. As with other deposits in N. America, there is, however, a high arsenic content, which together with a substantial pyrrhotite content (~75%) has acted as a disincentive for re-processing.

However, an environmentally-clean and efficient chloride-based approach to re-treating these tailings was proposed in 2010 (Dry & Harris, 2010), and more recently in 2020 (Harris & Dry, 2020), and also a bioleaching process has been suggested (Cameron et al., 2018; Sudbury Star, 2019). The chloride approach can recover, in addition to nickel, cobalt, copper and the PGMs, useful products of iron and sulfur, as well as the intrinsic energy contained in the tailings. Additionally, it fixes 100% of the contained arsenic as scorodite, widely recognized as being the most effective arsenic-fixation method. Bioleaching, whilst being able to recover nickel, copper and possibly cobalt, is not able to achieve anything else, and more importantly, will generate a tailings volume appreciably greater than that already there due to the oxidation of iron and sulfur to form ferric hydroxide and gypsum. Furthermore, bioleaching of pyrrhotite, especially a material with 75% pyrrhotite such as the tailings, is actually quite hazardous due to the huge amounts of heat generated, and bioleaching once it starts is very difficult to suppress. However, where there are no recoverable economic values from such tailings, the recent work on inhibition of sulfide oxidation may serve to reduce or even halt the oxidation of pyrite and pyrrhotite (Kalin et al., 2018). CPMW were applied in the field to fresh pyrrhotite tailings. Oxidation rates were estimated after outdoor exposure 3.2 years followed by 5.5 years of indoor storage. The leachate was monitored for 1.8 years.

Thus, in one fell swoop, the pressure could, at least partially, be alleviated on the supply side, and these old liabilities could be re-processed and remediated, ideally ecologically, to the benefit of the global population as well as the restoration of various types of landscapes. The industry, unfortunately, despite the many tailings dam disasters, seems very reluctant to adopt this approach.

Another potential new source of supply is the so-called deep-sea nodules, which contain vast quantities of manganese, nickel and cobalt, several times the identified terrestrial reserves. Interest in the potential exploitation of polymetallic seabed nodules generated a great deal of activity, interest in and excitement among prospective mining consortia in the 1960s and 1970s. Almost half a billion dollars was invested in identifying potential deposits and in research and development of technology for mining and processing the nodules (Wikipedia, 2019). These initial undertakings were carried out primarily by four multinational consortia composed of companies from the United States, Canada, the United Kingdom, the Federal Republic of Germany, Belgium, the Netherlands, Italy, Japan and two groups of private companies and agencies from France and Japan. There were also three publicly sponsored entities from the Soviet Union, India and China (Wikipedia, 2019).

In the late-seventies, two of the international joint ventures succeeded in collecting several hundred-tonne quantities of manganese nodules from the abyssal plains (18,000 feet, >5.5 km depth) of the eastern equatorial Pacific Ocean. Significant quantities of nickel (the primary target at the time) as well as copper and cobalt were subsequently extracted from this "ore" using both pyrometallurgical and hydrometallurgical methods.

An Economist report suggests that harvesting of deep-sea nodules is once again definitely back on the agenda (The Economist, 2017). However, any future mining of nodules will need to be authorized by the International Seabed Authority (ISA) and would need to quantify any impact in advance via an Environmental Impact Statement. This, of course, introduces a very contentious topic, namely as to whether deep sea mining should be allowed from an environmental perspective. There are, as might be expected, a diversity of opinions about the impact seabed mining might have on the local ecosystem, and the fact is that nobody really knows. One theory is that since the nodules are generally found in the proximity of hydrothermal vents, which are constantly active, then harvesting of nodules would not, in fact, cause any disruption different to that what is already there (The Economist, 2017).

A final, more far-out (literally) resource is space mining of asteroids. This clearly a long-way off, but one which is already being considered. Unlike with the deep sea, there are no environmental issues to confront (that we know of), so it maybe, if the technology can be developed quickly enough, that space mining might happen before deep-sea mining. We have about two decades to bring one of these into reality.

3.4 Modern Chloride Extraction

Despite a track record of "nonsuccess" of the many chloride-based processes for sulfide feeds, there remain many compelling reasons why the application of chloride chemistry not only can result in improved processing, but also can contribute greatly to achieving sustainability and improved environmental performance. There have been, and still are, a few very successful chloride-based base metal operations. Glencore (formerly Falconbridge) has operated a chloride process for many years at its nickel-cobalt refinery in Kristiansand, Norway, which is arguably the best base metal recovery plant in the world today. It was initially a hydrochloric acid leach, but more lately has used chlorine as the main oxidant/lixiviant, largely because chlorine is available, being generated from the subsequent electrowinning circuits for the recovery of nickel and cobalt metals (Stensholt et al., 1986a, 1986b, 1988; Thornhill et al., 1971). Noranda (as it then was) operated the Brenda Leach Process, which employed a high temperature (105°C to 110°C), high-strength chloride (30% $CaCl_2$ + NaCl + HCl) atmospheric leach of copper-molybdenum concentrates until the mine shut down in the 1990s (Jennings et al., 1973). This process was highly efficient, and essentially leached out all the copper, lead, and calcium from molybdenum concentrates to allow further, conventional processing of molybdenum to take place, without generating large quantities of toxic residues.

Most of the advantages that were originally expected from the use of chloride (Harris, 2014) with the processes developed in the 1970s and 1980s have remained and are briefly discussed below.

3.5 Leaching and Intrinsic Energy Content of Sulfides

Intrinsic energy can be substantially recovered, especially if the feed contains appreciable levels of pyrrhotite (Harris et al., 2007). The presence of pyrrhotite in an ore or concentrate of all the iron sulfide minerals is generally regarded as a major disadvantage, as it can be a significant factor in acid mine drainage. This is equally true for sulfate leaching, but the advantage that chloride has is that the acid used to effect leaching can be recovered and recycled (see iron below) which is not the case with sulfate. Processing sulfide minerals in this way to recover energy as heat is environmentally advantageous, since for every GJ of heat recovered, an equivalent amount of burning carbon is prevented, and a toxic, acid-generating waste is eliminated. Chloride circuits can be operated at atmospheric pressures, and are more readily adjusted to ensure that the sulfide-sulfur that accompanies the mineral either ends up as H_2S gas, or as elemental sulfur, a product that can also be sold, and both forms can be converted to sulfuric acid.

Chloride extraction circuits are more aggressive than their sulfate counterparts. This has the advantage that higher metal recovery can be achieved, along with a residue that is often easier to filter, which has a lower volume, and is less prone to metal/acid leaching into the environment. Indeed, most chloride leach residues are predominantly benign alumino-silicate gangue.

3.5.1   Iron and Hydrochloric Acid

Iron is the major contaminant in virtually every hydrometallurgical processing circuit, and has been deemed worthy of five international conferences devoted entirely to its control and disposal. Sulfate chemistry is such that iron must be precipitated via the use of some form of a base or neutralizing agent, generating large volumes of sludges, whether they be jarosite, goethite, hematite, or "ferric hydroxide." However, chloride chemistry affords the possibility not only of recovering the associated acid for reuse, but also of generating a marketable iron product, namely hematite. At the very worst, this hematite is easy to filter, has a low volume, and is environmentally benign.

3.5.2   Aluminum and Magnesium

Because of the highly aggressive nature of the chloride leaching operation, both aluminum and magnesium tend to report to the resultant leach filtrate in significant concentrations. During acid recovery, through hydrolysis of the iron chloride, the aluminum reports virtually 100% along with the hematite. However, the different crystal structures of the two oxides result in discrete compounds, allowing easy separation of the aluminum. Magnesium, on the other hand, remains in the liquid phase when either iron or aluminum is present, thus affording an efficient and simple separation. It can be recovered in a subsequent hydrolysis step as a magnesium oxychloride, which can be calcined and results in a marketable magnesia.

3.5.3   Environmental Aspects

Because chloride is so aggressive, as noted above, it tends to dissolve all metals from the ground rock. Thus, leach residues, which in conventional processing are generally voluminous and must be ponded as tailings, are low in volume, are generally crystalline, and, most importantly, are no longer reactive and hence are environmentally nonthreatening.

3.6   Current State of Development

Considerable development work has been undertaken on the chloride-based process over the past decade. Ideally, the mining and metals extraction industry will consider embracing what is essentially a quantum change in how it goes about its business, and at the same time will overcome the negative perceptions created in the past due to the many failed processing routes. The fact that microbial processes can be controlled is generally not understood, and hence not accepted, by the industry, largely because it is not within "normal" paradigms, which are based on the principles of classical inorganic chemistry. Such ecologically based technologies, which would reduce or completely halt the weathering rate at source, are generally either ignored or are declared to be impossible or uneconomic, unfortunately with no real basis for such declarations.

Further, without solid geomicrobiological knowledge, biological oxidation/corrosion control cannot be implemented. There is, nevertheless, in this respect an opportunity for the mining industry to embrace a technology (biological or ecological engineering) that could have far-reaching benefits. The industry needs to take the risk.

In the contexts of both efficiently recovering more value from the material mined, and at the same time appreciably reducing the amounts of toxic wastes generated, revisiting chloride processing as it was originally conceived in the 1970s has merit. In recent times, chloride-based extraction has been further developed and refined, with a greater understanding of the parameters involved, and with the objectives discussed above in mind, to the extent that it can now be considered a viable option.

References

Batty, J. D., & Rorke, G. V. (2006). Development and commercial demonstration of the BioCOP™ thermophile process. *Hydrometallurgy, 83*(1), 83–89.

Cameron, R., Yu, B., Baxter, C., Plugatyr, A., Lastra, R., Dai-Cin, M., Mercier, P. H. J., & Perreault, N. (2018). Extraction of Cobalt and Nickel from a Pyrrhotite Rich Tailings Sample via Bioleaching, Proceedings Extraction 2018 (Boyd R. Davis, Michael S. Moats and Shijie

Wang, Editors), Ottawa, Ontario, August 2018, The Minerals, Metals & Materials Series p. 2669, https://doi.org/10.1007/978-3-319-95022-8_225.

Dry, M. (2015). Technical & cost comparison of laterite treatment processes: Part 3. *Proceedings of Alta Ni/Co conference*, May 2015, Perth, WA, p. 24. Alta Metallurgical Services, Perth, WA. Retrieved from https://www.altamet.com.au/wp-content/uploads/2015/07/ALTA-2015-NCC-Proceedings-Contents-Abstracts.pdf

Dry, M and Harris, B. (2010). Nickeliferous Pyrrhotite – Another source of Nickel if it can be extracted economically, ALTA Ni-Co-Cu 2010, Perth, WA, May 24–29, 2010.

Flett, D. S., Melling, J., & Derry, R. (1983). Chloride hydrometallurgy for the treatment of complex sulfide ores. *Warren Spring Laboratory Report LR 461 (ME), U.K.*

Halbe, D. (2003). Business aspects and future technical outlook for hydrometallurgy. In C. Young, A. Alfantazi, C. Anderson, A. James, D. Dreisinger, & B. Harris (Eds.), *Hydrometallurgy 2003* (Vol. 2, p. 1091). The Minerals, Metals and Materials Society.

Harris, G. B. (2014). Making use of chloride chemistry for improved metals extraction processes. In *Proceedings of the 7th International symposium on hydrometallurgy* (HYDRO 2014) (pp. 171–184). Canadian Institute of Mining, Metallurgy and Petroleum.

Harris, G. B., & Dry, M. J. (2020). Re-treatment of tailings using chloride-based processing, Presented at *COM 2020 Virtual On-Line Conference, Proceedings of the 59th Conference of Metallurgists, COM 2020*: The Canadian Institute of Mining, Metallurgy and Petroleum. ISBN: 978-1-926872-47-6, October 2020. Also presented at *ALTA Ni-Co-Cu 2020 Virtual On-Line Conference*, November, 2020, 440–453.

Harris, G. B., White, C. W., Demopoulos, G. P., & Ballantyne, B. (2007). Recovery of Copper from a Massive Polymetallic Sulphide by High concentration Chloride Leaching, In Proceedings of *Copper 2007, the John E. Dutrizac Symposium on Copper Hydrometallurgy, Sixth International Copper-Cobre Conference*, Toronto, August 25–30, 2007.

Heikkinen, V., & Korte, M. (2019). Bioheapleaching in Boreal Conditions - Temperature Profile Inside the Heaps and Microbiology in Elevated Temperatures, ALTA Ni-Cu-Co 2019, Perth, WA, May 18–25, 2019.

Jennings, P. H., Stanley, R. H., & Ames, H. L. (1973). Development of a process for purifying molybdenite concentrates. In D. J. I. Evans (Ed.), *Proceedings of second international symposium on hydrometallurgy* (p. 868). American Institute of Mining, Metallurgical, and Petroleum Engineers.

Kalin, M., Wheeler, W. N., & Bellenberg, S. (2018). Acid Rock Drainage or not - oxidative vs. reductive biofilms—A microbial question? *Minerals, 8*(5), 199; Retrieved from:. https://doi.org/10.3390/min8050199

Kuhn, M. C., Arbiter, N., & Kling, H. (1974). Anaconda's Arbiter process for copper. *CIM Bulletin, 67*(742), 62–71.

McLean, D. C. (1982, February). Chloride leaching of copper concentrates: Practical operational aspects. Paper presented at 111th AIME Annual Meeting, Dallas, TX.

Neale, J., et al. (2016a). The Mondo Minerals Nickel Sulfide Bioleach Project: From Test Work to Design, Presented at *ALTA NCC 2015*, Perth, WA, May 21-25, 2016, p. 373.

Neale, J., et al. (2016b). The Mondo Minerals Nickel Sulfide Bioleach Project: Construction, Commissioning and Early Plant Operation, *ALTA NCC 2016*, Perth, WA, May 23–25, 2016.

Paynter, J. C. (1973). A review of copper hydrometallurgy. *Journal of the South African Institute of Mining and Metallurgy, 74*(4), 158–172.

Peek, E., Barnes, A., & Tuzun, A. (2011). Nickeliferous Pyrrhotite – "Waste or resource?". *Minerals Engineering, 24*, 625.

Royal Society of Chemistry. (2011). *A Sustainable Global Society. Chemical Sciences and Society Summit White Paper*. RSC.

Sackett, P. (2012). Endangered elements: Conserving the building blocks of life. *Solutions, 3*(3), June 2012. Available at https://www.thesolutionsjournal.com/article/endangered-elements-conserving-the-building-blocks-of-life/. Accessed 13 Aug 2019.

Stensholt, E. O., Zachariasen, H., & Lund, J. H. (1986a). Falconbridge chlorine leach process. *Transactions IMM, 5,* C10.

Stensholt, E. O., Zachariasen, H., & Lund, J. H. (1986b). The Falconbridge chlorine leach process. In E. Ozberk & H. Marcusson (Eds.), *Nickel Metallurgy, Volume I—Extraction and Refining of Nickel* (p. 442). CIM.

Stensholt, E. O., Zachariasen, H., Lund, J. H., & Thornhill, P. G. (1988). Recent improvements in the Falconbridge nickel refinery. In *Extractive metallurgy of nickel and cobalt*. TMS-AIME.

Sudbury Star. (2019). Sudbury's Mine Tailings Worth Billions, July 31, 2017. Available at https://www.thesudburystar.com/2017/07/31/sudburys-mine-tailings-worth-billions/wcm/4525f977-934d-5df3-96bc-5155a6fd230e. Accessed 19 Mar 2019.

The Economist (2017). Plucking minerals from the Seabed is Back on the Agenda, February 23, 2017. Available at https://www.economist.com/science-and-technology/2017/02/23/plucking-minerals-from-the-seabed-is-back-on-the-agenda. Accessed 13 Mar 2019.

Thornhill, P. G., Wigstøl, E., & Van Weert, G. (1971). The Falconbridge matte leach process. *Journal of Metals, 23*(7), 13–18.

USGS. (2019). Mineral Commodity Summaries, February 2019. Retrieved from https://prd-wret.s3-us-west-2.amazonaws.com/assets/palladium/production/atoms/files/mcs2019_all.pdf

Wadsworth, M. E. (1984). Trends and developments in copper metallurgy research. In *Copper '84 symposium. CIM 23rd annual conference of metallurgists*, Quebec City, August.

Wikipedia. (2019). Manganese nodules. Available at https://en.wikipedia.org/wiki/Manganese_nodule#Environmental_issues_and_sensitivities. Accessed 13 Mar 2019.

Chapter 4
Waste Management: A Brief History and the Present State

Margarete Kalin-Seidenfaden (iD)

Abstract In the past mines left their waste rock and tailings to weather, filling valleys, lakes and/or rivers. It wasn't until the end of WWII, that uranium mining wastes became a public concern. Mine waste remediation started gradually with erosion control and prevention of dust storms. Base metal mines started waste remediation gradually later, with the same objectives. Tailings surfaces of base-metal mines were stabilized with lime and grass covers. Some of these abandoned sites were invaded by native plants which were thought to transport radionuclides into the food chain, but no evidence of bioconcentration was found. However, acid and alkaline effluents remained of concern. Government, jointly with the mining industry, funded not only ecological inventories of tailings, but soon also programs addressing acid and alkaline contaminants in waste effluents. Boojum Research was funded within these programs to address these contaminants and provide decommissioning planning. Referring to comprehensive articles for readers interested in details, the chapter explains a main driving force for the improving waste management and effluent containment.

Keywords Environmental awareness · Sulfide backfilling · Radiation safety · Acid rock drainage · Reactive Acid Tailings Sulfide Program · National Uranium Tailings Program · Mine Environment Neutral Drainage · Food chain contamination · Constructed wetlands · Inorganic contaminants · Municipal waste

For centuries, mine wastes, were just that – wastes. Miners left waste rock and tailings to natural weathering processes, where atmospheric precipitation carried the weathering products through the wastes, creating acid mine drainage. Calculations based on weathering rates suggest that mine effluent contamination from many mines will continue for hundreds or thousands of years, since weathering occurs both under both aerobic and anaerobic conditions (Kalin & van Everdingen, 1988).

M. Kalin-Seidenfaden (✉)
Boojum Research Ltd., Toronto, ON, Canada
e-mail: margarete.kalin@utoronto.ca

© The Author(s), under exclusive license to Springer Nature Switzerland AG 2022
M. Kalin-Seidenfaden, W. N. Wheeler (eds.), *Mine Wastes and Water, Ecological Engineering and Metals Extraction*,
https://doi.org/10.1007/978-3-030-84651-0_4

Over time, our understanding of the origin and production of mine wastes has led to improvements in mine waste management. Several methods and technologies over the last century have emerged to lessen the impact of these wastes on the environment. Mine management practices have further evolved with the rise of environmental awareness. Table 4.1 summarizes mine waste management in the mining sector, comparing past and present (last 50 years) practices (Kalin, 2004). The comparisons of the past and the present reflect largely an astute awareness that mining wastes are presently confined to the mine waste management area, leaving a smaller footprint. In addition to this progress, the long-term generation of acid mine drainage is recognized through financial assurances since the early nineties, expressed well in an article in the Mining Journal entitled "No Simple Solution" (Knapp & Walsh, 1991). Financial assurances for perpetual treatment are the accepted solution for the decommissioning of a mine waste management area. These changes were initiated mainly in the uranium industry and have been translated to other mining operations. However, they have not necessarily been implemented worldwide. None of the waste management practices listed in Table 4.1 address the role of microbes in generating contaminants. Remediation strategies have been selected based on physical confinement and chemical reactions (neutralization reagents), relying upon retention of contaminants through the reduction or exclusion of oxygen, followed by application of neutralizing agents in water treatment plants.

Several practices have been implemented in the mining industry to reduce the environmental impact. The first practice is segregating the sulfides and backfilling underground workings with a high density paste made from tailings. Initially the

Table 4.1 Comparison of past and present mining waste management site selection and design for waste rock and tailings over the last 50 years

Present	Past
Site selection for waste rock and tailings with hydrological and economic considerations	Economic considerations only (e.g., proximity to mine)
Ore stockpiles placement and exposure	Not considered
Run-off drainage systems isolated from contaminated flows	Sometimes considered
Progressive reclamation of site during operations	Not considered
General mine-closure plan considered	Not considered
Strict design criteria for storage facilities for chemicals and fuel	Sometimes considered
Segregation and stockpiling of rock types according to acid-generating potential	Unsegregated waste rock piles
Improved dam design, including liners and leak detection systems	Dams constructed from coarse tailings, overburden, or waste rock
Thickened tailings; Underwater tailings management facilities	Above-ground tailings management facilities, no thickening
Tailings cleaning—sulfide separation	Not available
Tailings: high-density paste backfill	Not available
Highly acid-generating material used as backfill	Conventional backfilling, using only coarse fraction of the tailings

Source: Kalin (1998)

practice of backfilling high sulfide wastes was considered an environmentally desirable option, but experience in some Canadian mines demonstrated that the high-sulfide waste used for backfill could catch fire and prevent further mine operation. Thus, for safety reasons, the sulfur content of mine backfill is severely restricted. Generally, though, the volume of broken rock and tailings exceed the volume of the cavities created by mining. Hence, it is not possible for all generated wastes to be accommodated in the mine voids from which they were extracted. The surplus must be stockpiled, unless a use can be found for it as aggregate (sand or gravel), if the sulfur content is negligible. The challenge of isolating or otherwise finding beneficial uses for waste rocks and tailings from open pit operations remains.

The overburden and waste rock from open pit operations must be stockpiled outside the pit during active pit operation and can comprise up to ten times the ore volume. Returning this waste material to the pit when it is mined out is generally prohibitively expensive. Sometimes pit benches are filled with rocks allowing a high hydraulic conductivity. This allows groundwater to flow along the pit walls rather then through the tailings. It is called surround grout construction. Thus, when the pit is mined out, where possible, the void is filled with water by force flooding or allowing the groundwater and rain to gradually fill, creating pit lakes. In Germany, the former coal mining voids (open cast mines) were filled with river water (Jordans, 2018), although not all operations have been successful, as iron-laden water has emerged in the Spree river (IGB [Leibnitz Institut für Gewässerökologie und Binnenfischerei], 2018).

There are companies that segregate sulfides in tailings and store them under water, pending the day when they can be economically processed. It is also common practice to use and isolate waste rock and tailings as backfill in underground mining operations. Tailings-paste fill operations use the non-sulfide fraction of the tailings backfill after thickening, allowing immediate recycling of water. Sand-fill operations have the option of thickening the slime portion at the concentrate level, producing a thickened product as a valuable, impermeable cover for old tailings and waste rock deposits. Both processes allow the immediate recycling of water, a useful measure.

4.1 Mine Waste Site Ecology: The Beginning and Food Chain Contamination

The generally accepted restoration technique for mining wastes applies lime and fertilizer, followed by crimping of straw (GARD Guide; Verburg et al., 2009). This is usually followed by seeding with a commercial grass seed/legume mixture. The reclamation of uranium tailings in Canada followed the same methodology. In some early mines in the Northwest Territories of Canada, though, the tailings areas were left for indigenous species to repopulate.

The roots of naturally invading trees and shrubs were likely to penetrate deeper than the roots of the grass and legume covers, concentrating toxic metals from the tailings in their tissues. This led governmental regulators and scientists to raise concerns over potential food chain contamination through this indigenous vegetation.

Of serious concern were the long-lived radionuclides contained in the uranium mining wastes.

The Institute of Environmental Studies (IES) at the University of Toronto, Ontario, Canada launched investigations of the indigenous flora on alkaline, barren uranium tailings abandoned for 10 to 15 years. A diverse flora of indigenous, terrestrial, and aquatic biota was found (Kalin, 1984). A total of 15 uranium tailings sites were studied, both acidic and alkaline, and re- and un-vegetated, in the province of Ontario, Canada. Later the uranium mine sites in the Northwest Territories and the Province of Saskatchewan (Kalin, 1985) were included. Another part of the funding supported an MSc thesis (Caza, 1983) to study the growth and colonization of Trembling Aspen on uranium tailings (Fig. 4.1a).

Radium-^{226}, Uranium-^{236}, and Lead-^{210} concentrations were determined in both terrestrial and aquatic vegetation as well as in the tailings around the root areas. As a result of this work, it became evident that indigenous terrestrial plants posed no threat to the food chain, as the radionuclides and most metals remained generally in the roots and surrounding soils. Tree roots form a dense carpet-like structure below

Fig. 4.1 (a) Trembling Aspen stand growing on bare tailings. (b) A root carpet is lifted to document the horizontal root growth of the grass cover. (c) Root penetration of the grass cover showing gray and brown regions. Gray areas are unoxidized tailings. (Photographs by M. Kalin-Seidenfaden)

the grass cover on seeded tailings (Fig. 4.1b), surrounded by iron precipitate (Figs. 4.1b, c). The vegetated surface covers reduced wind dispersal and erosion of the tailings, while decreasing rainwater infiltration. Figure 4.1c highlights the difference between oxidized and unoxidized areas of the tailings. The oxidized areas (brown) contained elevated levels of radionuclides.

4.2   Boojum and Government/Industry Programs

The uranium industry was one of the first to include environmental issues in their close-out plans, as public awareness raised these issues in the late 1960s. From these efforts, several principles were developed to govern management practices, such as ALARA (As Low as Reasonably Achievable) for radiation safety at uranium operations. It was later followed by BATEA (Best Available Technology Economically Achievable) for all other mining operations (Pouw et al., 2015). Comprehensive historical reviews of risks and environmental policy have been written by Faber & Wagenhals (1988) and Kamieniecki & Kraft (2013). These efforts are commendable and have brought about significant change in the mining industry.

The accepted treatment of contaminated drainages from both tailings and waste rock piles has remained the same for decades. Neutralizing agents, such as lime, are added to acid streams leaving a metal-laden sludge behind which needs further stabilization. The reactivity of the sludge depends on the pH of effluent, the lower the pH the greater the reactivity (McDonald et al., 2006). It is often returned to the tailings piles. The neutralisation leads to an increase in pH with the formation of inorganic particulates, which settle out of the water column, either with time or supported by flocculating agents. With aging of the sludge, and through microbial activity, the metals are released again. A research team at NRCan (Natural Resources Canada) addressed the stability of the resulting neutralizing sludge and concluded: "Current sludge management practices are ad hoc and frequently do not address long-term storage" (Zinck, 2006).

INAP (International Network for Acid Prevention) has created a guideline, the GARD Guide (Global Acid Rock Drainage: GARD), which is an internationally recognized guide to the prediction, prevention, and management of drainage produced from sulfide mineral oxidation (Verburg et al., 2009; Kleinmann & Chatwin, 2011). In accordance with the guide, most current mine operation practices emphasize containment of the wastes, thereby reducing the volume of effluent, not its quality. These containment practices require significant financial commitment from the operating mining company. But, while the management practices outlined in the GARD Guide certainly reduce the immediate environmental impacts, they may, in many ways, delay the onset of longer-term mine drainage issues. These entrenched practices are a hindrance to novel approaches to mine waste management and the acid challenge.

Remediation efforts and drainage treatment are viewed by some in industry and government as 'the price to be paid' and therefore accepted as part of mining and metal extraction. To some degree, physical and chemical aspects of natural weathering processes are abated by present mining practices, but the fundamental contribution of microbial populations is ignored. Herein lies the long-term challenge. Only when the microbial oxidation is controlled will long-term weathering processes subside. Hence, current mine environmental management practices are, in a true sense, not sustainable.

While discussions on the food chain were ongoing, some regulators were starting to think about the idea of declaring mine wastes as hazardous materials. Mine wastes are rocks, broken or ground, and certainly not hazardous. They need appropriate handling, as rocks are part of nature, supplying essential elements supporting living systems in the aquatic, terrestrial and even in the atmospheric areas of the planet. The challenge arises due to the very large surface area of mineralized rock that is exposed. The weathering of this rock releases an excess of some elements which, in many cases are toxic to the surrounding ecosystems, altering the pH and the electrical conductivity- two key drivers of ecosystem change.

As food chain contamination was no longer a pressing issue, concerns turned to contaminated fresh- and groundwater. The uranium industry in Canada anticipated the development of more stringent environmental regulations, and was seeking sustainable, ecological approaches to address drainages from their wastes and for decommissioning of mine waste management areas. The Canadian government then funded a 5-year program, the National Uranium Tailings Program (NUTP), in 1981, to address the long-term environmental impact of uranium tailings. The long-term goal for these was to seek a sustainable approach to the decommissioning of mine waste management areas. With the encouragement of the government and the uranium industry, Boojum Research Ltd. (Boojum) was founded in 1982 as an R&D company. Its objective was to find long-term, sustainable, economic solutions to mine closures.

Boojum Research's first assignment under NUTP was to remove contaminants from alkaline uranium mine waste holding ponds. Several abandoned, pH-neutral, tailings ponds were investigated for their indigenous aquatic floras. In base metal and gold tailings, extensive meadows of *Chara vulgaris* were found growing (Figs. 4.2a, b). These algae appeared to be ideal biological polishers, as they were also fast growing, and did not transport contaminants from the sediment back into the water when the biomass decayed, but relegated the biomass and contaminants into the sediment.

Since phosphate is often limiting to aquatic plant growth, supplementing phosphate to *Chara was* investigated to alleviate one of the forcing functions restricting their growth and productivity. The effects of phosphate on the growth of Chara supported by NRC IIPAP funds, the Masters thesis of M. P. Smith (1988). Forcing functions are defined as one or more resources that halt or slow progression of further development (see Chapter 5 for details). This research was supported by Boojum and an IRAP grant (Industrial Research Assistance Program of the National Research Council of Canada) and lead to introduction of *Chara* as bio-polishers to ponds where they did not previously exist. After several failures, the algae were finally established in several tailings ponds, supporting metal and radionuclide removal (Kalin & Smith, 1986).

Fig. 4.2 (**a**) Boojum researcher, M.P. Smith, holding a sample of *Chara vulgaris* found in a nickel mine tailings pond. (**b**) Gold mine tailings pond in northern Ontario with dense populations of *Chara vulgaris* (in the ponds). (Photographs by Boojum Research)

The NUTP research program was followed in 1983 by the Reactive Acid Tailings Sulfide Program (RATS). This program focused on modeling, prediction, and methodologies to reduce or remediate the long-term environmental effects of acid-generating materials (John & Joe, 1987). Boojum's first project under the RATS program took place on a tailings site covered with a hard-oxidized crust of pyrrhotite (FeS). When unoxidized pyrrhotite is exposed to moisture, it starts to burn. Rains produced acid run-off. Further, the mining company could not risk using heavy re-vegetation equipment to establish a cover, as the crust could not carry the weight of the heavy equipment. The crust would break, exposing the un-oxidized FeS, and rain or moisture would ignite it.

Field surveys showed that cattails (*Typha sp.*), moss, horsetails (*Equisitum sp.*), and blue-green algae (cyanophytes) grew along the banks of a nearby alkaline mine slime stream. The alkaline stream resulted from washing explosives from the underground walls or using shotcrete to cover acid-generating walls (Jones & Wong, 1994). Similar organisms were found growing on the tailings, mostly associated with sticks or rocks, despite the reactivity of the pyrrhotite. On the edge of the barren tailings crust, vegetation was noted, similar to that along the alkaline creek. The distribution patterns were comparable to those observed on uranium tailings several years earlier (Kalin, 1984). Detailed measurements in the colonized areas generally produced hints that physical topography (rocks providing shade or decaying wood) and/or chemical conditions were growth-supporting factors.

The FeS tailings needed to be covered to reduce the acid run-off. The first level of 'vegetation' included a moss cover as it could colonize bare surfaces. Early attempts to grow moss in greenhouses and onsite were unsuccessful, but with the proper fertilizer and shade, a green haze developed in some of the boxes in the greenhouse trials (Fig. 4.3a). The treatment that produced the greenest boxes was translated to the field, where trials were run to determine the best season for starting (Fig. 4.3b). One example of successful plant growth was an island overgrown with

Fig. 4.3 Ecological engineering measures for stabilization and cover for pyrrhotite tailings. (**a**) Tailings covered with alkaline mine slimes fosters colonization by indigenous plants. (**b**) Experiments to fertilize the pyrrhotite tailings surface at various times during the season. Fall fertilization was successful. (**c**) Horsetails colonized the alkaline mine slimes without any fertilizer. In background are the permeable waste rock dikes to accumulate mine slimes. (**d**) Tailings several years after decommissioning recommendations implemented. (Photographs by Boojum Research)

horsetails (*Equisitum sp.;* Fig. 4.3c). This overgrowth had established without our help and served as an example for our trials. At the end of the RATS program, we recommended to the mine operators that they create permeable dikes (composed of coarse, un-compacted, larger waste rocks) and divert mine slime streams into these dikes. The mine manager implemented the recommendations and after several growing seasons, native horsetails and moss covered the site. A photo taken 10 years later shows the success of the treatment (Fig. 4.3d).

In addition to the alkaline stream on the pyrrhotite tailings, a slow-moving, acidic (pH 2.5) creek was chosen on the site to address the forcing functions for aquatic acid systems (Fig. 4.4a). The creek water contained high concentrations of sulfate (4–6 g.L^{-1} SO$_4$) and dissolved iron (1–2 g.L^{-1}). Loose straw (not bales) was used as an organic carbon amendment for microbial growth. The straw was added to several sections of the creek (Fig. 4.4b). In the winter, while drilling holes in the ice cover, hydrogen sulfide was released. In the spring, clear water was found in the straw-filled section. Within the straw the pH was up to about 3.5, a remarkable increase from the low of 2.5 (Fig. 4.4c).

The improvements in the creek water had been clearly induced by microbial activity. The ice cover on the creek slowed the flow, while reducing wind-driven mixing and oxygen diffusion. Heterotrophic microbes growing on the straw

Fig. 4.4 (a) Pyrrhotite tailings crust. In the foreground is the creek in which the first straw addition was made. (b) Straw addition in acidic creek before winter at the time of setup. (c) Clear section of creek with straw in spring, all iron had been precipitated. (Photographs by Boojum Research)

consumed oxygen, lowering the redox potential of the water. The combination of low oxygen and organic carbon fostered the growth of anaerobic, iron- and sulfate-reducing microbes. Stumm & Morgan (1996; p. 477) provided a clear explanation for geomicrobiology processes. In general, a group of heterotrophic microbes alters the surrounding growth conditions by making them a little less oxidative. This, in turn, provides the proper conditions for the next microbe group to lower the redox state even further. When the local conditions are reducing, microbes such as iron and sulfate-reducing microbes will precipitate iron onto the straw while increasing the pH. The emerging smell of H_2S, the rotten egg smell noted in the winter suggested that not enough oxidized iron had remained in the water to form iron sulfide. If the pH had been high enough and the E_h low enough some pyrite could possibly have formed (Fernández-Remolar et al., 2003; Reitner & Thiel, 2011).

These experiments provided key observations for mine waste and water management, such as:

- No microbes were needed to seed the acidic water; they invade or awake when food is available.
- Iron precipitate covered the straw, reducing access to the organic carbon, not desirable.
- Ice cover reduced oxygen access, giving anaerobic microbes a chance to flourish.

Iron reduction by microbes raises pH, and this, in turn, leads to the in-situ metal precipitation. To reproduce these conditions in any mine effluent, two things needed to be developed. First, a living, floating vegetation cover would replace the ice cover. This would provide a continuous supply of organics, through decomposing litter and root exudates, and it would also decrease wind mixing of the water. Second, an iron-precipitation pond is needed upstream of the living cover to prevent intense iron encrustation of the root systems. In the creek, the straw became encrusted with iron, forming secondary mineral spheres (Fig. 4.5). The straw

Fig. 4.5. A piece of straw under the dissecting microscope from the creek to which straw was added. Note the iron precipitates (mineral balls) on the straw. (Photographs by Boojum Research)

provided organic carbon supporting the establishment of oxygen-consuming microbes. The microbial-based treatment system thus developed was named Acid Reduction Using Microbiology or ARUM.

The RATS program was followed by the Mine Environment Neutral Drainage (MEND) program in 1989. All programs were under the auspices CANMET (Canada Centre for Mineral and Energy Technology) of the NRC (Natural Resources Canada). MEND funding supported projects developing constructed wetlands for the treatment of AMD. Officials of these programs expected the same successes found while treating organic waste waters. They anticipated that same processes would take place with inorganic substances (Kadlec & Knight, 1996; Mitsch & Gosselink, 2000).

The stated objective for MEND, as Boojum understood it, was to "...develop technologies to prevent and control acidic drainage, or -- how to stop the lime trucks." These technologies were to address decommissioning of mine waste and water management areas, working within the wastewater management areas, before discharge to the receiving environment. The technologies to be developed were to provide acceptable conditions for a sustainable, walk-away so that the treatment plant could eventually shut down.

Although Boojum Research obtained funding under all government/industry programs it continued under MEND, but eventually differences in objectives lead to a divergence. Boojum Research focused on containing the weathering products within the waste and water management area (In Chapters 8 and 9 detail), whereas MEND reviewers were seeking solutions in wetlands (Kadlec & Knight, 1996).

The construction of microbially-active sediments and the addition of targeted nutrients to reduce or deactivate the predominantly oxidative environment within wastes is the focus of our ecological engineering solutions. The goal was the development of ecological tools which would improve the drainage leaving the site. Hence, Boojum's challenge was to determine the forcing functions within the mine wastes, alleviate them, and have natural bio-geochemical processes transfer contaminants to sediments and transform them back into ore bodies of the future (Debus, 1990).

References

Caza, C. (1983). "Biology of *P. tremuloides* on abandoned Uranium mill tailings sites near Bancroft, Ontario". M.Sc. Thesis, University of Toronto, Department of Botany.

Debus, K. (1990). Mining with microbes. *Technology Review, 93*(6), 50–57.

Faber, M., & Wagenhals, G. (1988). Towards a long-term balance between economics and environmental protection. In W. Salmons & U. Förstner (Eds.), *Environmental management of solid waste* (pp. 227–242). Springer.

Fernández-Remolar, D. C., Rodriguez, N., Gómez, F., & Amils, R. (2003). Geological record of an acidic environment driven by iron hydrochemistry: The Tinto River system. *Journal of Geophysical Research: Planets, 108*(E7).

IGB. (2018). Sulfate in River Spree and Lake Müggelsee. https://www.igb-berlin.de/en/project/sulfate-river-spree-and-lake-Muggelsee

John, R., & Joe, E. (1987). CANMET's Tailings Research Programs—An update. In *Proceedings of the 11th annual British Columbia mine reclamation symposium in Campbell River, BC, The Technical and Research Committee on Reclamation* (pp. 105–115). Available online.

Jones, C. E., & Wong, J. Y. (1994). Shotcrete as a cementitious cover for acid generating waste rock piles. In *Proceedings of the international land reclamation and mine drainage conference and 3rd international conference on the Abatement of Acidic Drainage* (Vol. 24, pp. 104–112).

Jordans, F. (2018). Germany turns former coal mines into vast lakeside resorts. *The Durango Herald, 6*, 22–2018. https://durangoherald.com/articles/229327

Kadlec, R. H., & Knight, R. L. (1996). *Treatment wetlands.* CRC Lewis Publisher sop 881, ISBN 0-87371-930-1.

Kalin, M. (1984). Port Radium, Northwest Territories: An evaluation of environmental effects of the uranium and silver tailings. University of Toronto, Institute for Environmental Studies. Retrieved from https://zone.biblio.laurentian.ca/handle/10219/3016

Kalin, M. (1989). Ecological engineering and biological polishing: methods to economize waste management in hard rock mining. In W.J. Mitch & S.E. Jorgensen (Eds.), *Ecological engineering* (pp. 443–461). wiley & Sons. ISBN 0-471-62559-0

Kalin, M. (1998). The Role of Applied Biotechnology in Decommissioning Mining Operations. Proceedings of the 30th Annual Meeting of the Canadian Mineral Processors, Ottawa, January 20–22 (pp. 154–167).

Kalin, M. (2004). Slow progress in controlling acid mine drainage (AMD): A perspective and a new approach. *Peckiana, Staatliches Museum für Naturkunde Görlitz, 3*, 101–112.

Kalin, M., & Smith, M. P. (1986). Biological polishing agents for mill wastewater. An example: *Chara.* In R. W. Lawrence, R. M. R. Branion, & H. G. Ebner (Eds.), *Fundamental and applied biohydrometallurgy* (p. 491). Elsevier.

Kalin, M., & van Everdingen, R.O. (1988). Ecological Engineering: Biological and geochemical aspects. Phase I experiments, In W. Salomons & U. Foerstner (Eds.), *Environmental management of solid waste* (pp. 114–128). Springer-Verlag. ISBN 3-540-18232-2.

Kamieniecki, S., & Kraft, M. (2013). *The Oxford handbook of US environmental policy.* Oxford University Press. Retrieved from https://global.oup.com/academic/product/the-oxford-handbook-of-us-environmental-policy-9780199744671?cc=ch&lang=en&#

Kleinmann, R. L., & Chatwin, T. (2011). The *GARD Guide* and its general applicability to mine water issues. In *Proceedings, American Society of mining and reclamation, Bismarck, North Dakota* Reclamation: Sciences Leading to Success June 11–16 (pp. 317–325).

Knapp, R., & Walsh, D. (1991). No simple solution, *CIM Bulletin*, June 1991, 63–66.

McDonald, D. M., Webb, J. A., & Taylor, J. (2006). Chemical stability of acid rock drainage treatment sludge and implications for sludge management. *Environmental Science and Technology, 40*(6), 1984–1990. https://doi.org/10.1021/es0515194

Mitsch, W. J., & Gosselink, J. G. (2000). Wetlands, 3rd Edition (p. 920).

Reitner, J., & Thiel, V. (Eds.). (2011). *Encyclopedia of Geobiology* (Encyclopedia of earth science series). Springer.

Pouw, K., Campbell, K., & Babel, L. (2015). Best Available Technologies Economically Achievable to manage effluent from mines in Canada. In *10th International Conference on Acid Rock Drainage and IMWA Annual Conference* (pp. 1–10).

Smith, M. P. (1988). Phosphorus Nutrition of *Chara vulgaris L.* MSc Thesis University of Toronto, 69 pp.

Stumm, W., & Morgan, J. (1996). *Aquatic chemistry: Chemical equilibria and rates in natural waters* (3rd ed.). Wiley.

Verburg, R., Bezuidenhout, N., Chatwin, T., & Ferguson, K. (2009). The global acid rock drainage guide (GARD Guide). *Mine Water and the Environment, 28*(4), 305.

Zinck, J. (2006). Disposal, reprocessing and reuse options for acidic drainage treatment sludge. In *7th International Conference on Acid Rock Drainage (ICARD)* (pp. 2604–2617).

Chapter 5
Constructed Wetlands and the Ecology of Extreme Ecosystems

Margarete Kalin-Seidenfaden (iD)

Abstract Constructed wetlands have been used for decades for the removal of organic pollutants. Organic contaminants are either degraded into air, gases or sequestered by aquatic vegetation. Inorganic elements need to be precipitated or adsorbed onto particulates, carried out of the water column, and stabilized in the sediments. Precipitation of these inorganic contaminants alters the hydrological conditions. This leads to plugging and hydrological changes with concomitant treatment failure, contrary to organic contaminants which are removed as gases or taken up by plant. Both processes are referred to as bioremediation, but the removal processes differ.

Mine sites should be treated as extreme ecosystems, surrounded by intact, native ecosystems with their seed sources. This unconventional view of approaching ecosystem development in areas of ground or broken rocks, such as deserts, hot springs, salt lakes, high latitude and altitude systems requires an understanding of the blockages which limit further progression. Mostly, mine waste systems need to be balanced, as they are overwhelmingly oxidative and either acidic or alkaline. Their special biogeochemical challenges present their own complexity, and they need to be given time.

Keywords Ecotechnology · Biogeotechnology · Ecosystem restoration · Ecological engineering · Niche construction · Biofilms · Geomicrobiology

5.1 Constructed Wetlands

Wetlands have gradually vanished world-wide as they are replaced by agricultural activities and urban development (Roy et al., 2000; Walpole & Davidson, 2018). The consequences of draining them were realized when their disappearance caused

M. Kalin-Seidenfaden (✉)
Boojum Research Ltd., Toronto, ON, Canada
e-mail: margarete.kalin@utoronto.ca

M. Kalin-Seidenfaden, W. N. Wheeler (eds.), *Mine Wastes and Water,*
Ecological Engineering and Metals Extraction,
https://doi.org/10.1007/978-3-030-84651-0_5

hydrological problems. The National Research Council (1992), in their report *Restoration of Aquatic Ecosystems* defined restoration as the "return of an ecosystem to a close approximation of its condition prior to disturbance." Further, they considered the resurrection of wetlands "the construction of a wetland in an area that was not a wetland in the recent past (within the last 100–200 years) and that is now isolated from existing wetlands (i.e., not directly adjacent)."

Organic pollutants are degraded into gases and nutrients, and support aquatic, above-ground and below-ground plant growth. They generate organic sediments which, in turn, house microbes which drive contaminant degradation resulting in changes in BOD (Biological Oxygen Demand). This provides reducing conditions through microbial action, gradually forming sediment.

The state of Florida initiated legislation to counter wetland losses using organic-rich wastewater and stormwater runoff -assisting the growth of wetlands. The restoration of the Everglades water quality remains one of the largest restoration efforts, reducing and altering the nutrient input from agricultural land in its headwaters (US EPA, 2016). This project, started in the 1980s, is still in progress restoring a previously straightened river which has caused dramatic damage to the Everglades system. Overall, these constructed wetlands perform well at their designated job, making available intensive research funding which resulted in increased understanding of the cleansing capacity of organics (Kadlec & Knight, 1996; Gwin et al., 1999; Mitsch & Gossilink, 2000), to name some of the most prominent publications.

Constructed wetlands were also designed to remediate soils, especially when contaminated with complex, man-made organics. The design of these wetlands was based on the role of microbes in degrading pollutants (Adriano et al., 1999). Organic man-made substances are complex, but their degradation and removal from wetlands by vegetation led to the development of the field of phytoremediation, addressing the transformation and control of a large variety of both natural and manmade contaminants (McCutcheon & Schnorr, 2003).

5.2   Ecology of Extreme Ecosystems

All these developments in constructed wetlands occurred long after nuclear weapons testing in the 1950s. It was also a period when acid rain and contamination with radionuclides from fall-out were of major concern. The first to note public pressure because of radiation and the potential food chain contamination was the uranium industry. The base metal industry followed as acid rain, resulting from tailings dust storms, became an issue. In northern Ontario, methods were developed for placing a grass cover on acid tailings, a standard which led to greening of base metal tailings (Peters, 1995) followed by uranium tailings. In the case of unattended orphaned sites, revegetation by indigenous flora took hold, but the effluents or drainages, acid or alkaline, remained contaminated with long-lived radionuclides. Hence public pressure increased on the uranium industry to assess the fate of ^{226}Ra and ^{210}Pb (see Chap. 4).

In 2002, the comprehensive book by Brown et al. (2002) on mine water treatment technologies published by the IWA (International Water Association) summarized the issue of mine water effluents very well along with approaches using various chemical treatments. These approaches were intensely debated at conferences, but never contained microbial or ecological engineering tools. However, Brown et al. (2002) quoted McGuinnes et al. (1996) recognizing ecological engineering with these words.

> All to often, constructed wetlands or other biological passive treatment systems are designed and constructed by engineers who do not have this perception and thus the systems do not achieve a self-sustaining status.

Bradshaw and Chadwick (1980) and Cairns (1980) were among the first to address the ecology and remediation of derelict and degraded lands. Their work focused on establishing a vegetation cover on coal spoils, gold, and base metal tailings. At the time, the accepted restoration technique applied lime and fertilizer, followed by crimping of straw followed by seeding with commercial grass seed/legume mixtures and later planting trees. The remediation of uranium tailings followed the same methodology.

Since constructed wetlands could and had worked well with organic pollutants, there was enthusiasm for possibly using constructed wetlands to treat mine effluents, as well. These effluents are either acidic or alkaline, but contain inorganic contaminants. Early attempts at using constructed wetlands for removal of metals worked for a limited time. However, the precipitation of metals driven by pH/E_h redox pairs (neutral pH and positive redox), soon 'plugged' the wetlands, with concomitant loss of removal capacity.

From the above discussion it would seem that the bioremediation of organic substances leading to gases, nutrients and water is a perfect tool, but mine tailings and their effluents contaminated by inorganic metals have a low pH and negative redox values. These effluents can only be treated by using a microbially-active sediment, which can raise the pH, precipitate metals, and keep them out of the effluent solution (see Chap. 6). The fundamental difference between constructed wetlands for organics removal and inorganics removal is the end product, either gas and nutrients, or precipitated solid metals from the inorganics. To use constructed wetlands for inorganics, there needs to be an organic, reducing sediment and microbes that can transform the redox state of the effluents. Boojum has focused its efforts on a containing the weathering products <u>within the waste by supporting the growth of organic microbial films</u> on the mineral surface and drainage effluents by creating organic, reducing sediments.

Mine wastes are not hazardous materials, but broken or ground rock, natural material exposed to natural weathering processes. Hence, ecological processes prevail in these wastes. Cairns' (1980) observations on the colonization of the dry coal spoils fitted reasonably well with of ecosystem processes and their development, as defined by Odum (1962). Odum formulated a restoration approach using 'ecological engineering' tools. He stated that "environmental manipulation by man using small amounts of energy to control systems" is all that should be required. This approach was later consolidated by Mitsch and Jorgensen (1989) into 'ecotechnol-ogy,' with 13 guiding principles, of which three apply specially to mining waste sites, as they are extreme, relatively simple ecosystems, with aquatic and terrestrial components.

1. Ecosystem structure and function are determined by forcing functions (missing resources) of the system. Alteration of these causes the most dras-tic changes in the ecosystem.
2. Ecosystems are self-designing systems. The more one works with the self-design of nature, the lower the cost of energy to maintain that system.
3. Elements are recycled in ecosystems. Matching humanity and the natural ecosystems supporting the biogeochemical pathways will ultimately reduce the effect of pollution and lead to a sustaining system. (Mitsch & Jorgensen, 1989; Chapter 3, pp. 21–25)

These three principles are fundamental to extreme ecosystems and therefore under-pin Boojum's approach to working in mine waste and water management areas. Only when the forcing functions are identified and dealt with, can the ecological engineering tools assist with the remediation of mine waste ecosystems and support the progression toward sustaining and self-supporting systems (Orellana et al., 2018).

Boojum has concentrated on the 'ecological engineering' tools that modify these extreme ecosystems to overcome the forcing functions which hold back any prog-ress toward sustainability. The tools include using local biota such as microbes, fungi, algae, mosses, and floating vegetation to alter the chemistry of the wastewa-ter. To eliminate forcing functions, organics and/or nutrients are added to the wastes supporting biofilm growth on the mineral surface or constructing sediments in drainage channels to establish reducing conditions. In some cases, physical altera-tions provide growth spaces for biota. In other words, Boojum has focused its work on determining which forcing functions limit the productivity of these extreme eco-systems and how they can be alleviated.

The first edition of the book on Ecological Engineering contained the observa-tions of a base metal mine with acid mine drainage (Kalin, 1989). Since 1989, Boojum has developed ecological and engineering tools intended to strengthen the drive toward mine site ecosystem sustainability. The concept of treating mine sites as extreme environments is fundamental to achieving the needed paradigm shift as demonstrated by the work produced by Boojum. Microbes and their productivity are the key ingredients enabling in the long-term, sustainably reduced weathering (Feldman, 1997).

Working with microbes in mining is nothing new. The book *Biogeotechnology of Metals* (Karavaiko & Groudev, 1985) opened with this quote from Pasteur (reprinted in 1969). The meaning remains valid to this date.

> There is no such thing as a special category of science called applied science; there is sci-ence and its applications, which are related to one another as the fruit is related to the tree that has borne it.

Working world-wide, in the high artic on Baffin Island in the Northwest Territories, the Yukon, Saskatchewan, Ontario, Quebec, Nova Scotia, Newfoundland (Canada), in the Minas Gerais (a tropical and in semi-arid region of Brazil, the Sahel of Burkina Faso, the humid subtropical region of Guiyang in China, the Rocky Mountains, West Virginia (USA), Northern Queensland in Australia, and in Germany convinced Boojum that, with minor differences, the same ecological principles persist at all mines. Generally, the forcing functions (lack of organic carbon, targeted nutrients, and physical settings) apply in all types of wastes sites, enabling the application of a standard approach to each site, like standard engineering practices.

Simply stated, Boojum applies natural processes found in extreme ecosystems to mine waste and water management. Many refereed publications and book chapters are publicly available, in addition to the gray literature which contains corporate reports of those companies which supported the development of decommissioning scenarios. The corporate reports and conference proceedings have been submitted in electronic form to the Laurentian University's Mining Environment library (https://zone.biblio.laurentian.ca/boojum).

References

Adriano, D. C., Bollag, J. M., Frankenberger, W. T., & Sims, R. C. (Eds.). (1999). *Bioremediation of contaminated soils* (American society of agronomy #37). Madison.

Bradshaw, A. D., & Chadwick, M. J. (1980). *The restoration of land: The ecology and reclamation of derelict and degraded land.* University of California Press. ISBN-13: 978-0520039612.

Brown, M., Barley, B., & Wood, H. (2002). *Minewater treatment: Technology, application and policy.* IWA Publishing. ISBN 1 84339 004 3.

Cairns, J. (Ed.). (1980). *The recovery process in damaged ecosystems.* Mich Ann Arbor Science Publishers. https://trove.nla.gov.au/work/9822333

Feldman, M. (1997). Geomicrobiological processes in the subsurface. A tribute to Johannes Neher's work. *FEMS Microbiology Reviews, 20*(3–4), 181–189. https://doi.org/10.1111/j.1574-6976.1997.tb00307.x

Gwin, S. E., Kentula, M. E., & Shaffer, P. W. (1999). Evaluating the effects of wetland regulation through hydrogeomorphic classification and landscape profiles. *Wetlands, 19*(3), 477–489.

Kadlec, R. H., & Knight, R. L. (1996). *Treatment wetlands* (p. 893). Lewis Publishers.

Kalin, M. (1989). Ecological engineering and biological polishing. In W. K. Mitsch & S. E. Jorgensen (Eds.), *Ecological engineering: An introduction to Ecotechnology* (pp. 443–461). Wiley.

Karavaiko, G. I., & Groudev, S. N. (1985). *Biogeotechnology of metals.* International training course on microbial hydrometallurgy of metals from ores (1982: Sofia, Bulgaria); International seminar on modern aspects of microbial hydrometallurgy (1982: Moscow, RSFSR). Centre of International Projects GKNT.

McCutcheon, S. C., & Schnorr, J. L. (2003). *Phytoremediation: Transformation and control of contaminants* (Wiley-interscience text and monographs). Wiley. ISBN: 0-471-39435-1.

Mitsch, W. J., & Gossilink, J. G. (2000). The value of wetlands: Importance of scale and landscape setting. *Ecological Economics, 35*(1), 25–33. https://doi.org/10.1016/S0921-8009(00)00165-8

Mitsch, W. J., & Jorgensen, S. E. (1989). *Ecological engineering – An introduction to Ecotechnology* (p. 472). Wiley.

National Research Council. (1992). *Restoration of aquatic ecosystems. science and technology and public policy.* The National Academies Press.

Odum, H. T. (1962). Man in the ecosystem. *Bulletin of the Connecticut Agricultural Station, 652,* 57–75.

Orellana, R., Macaya, C., Bravo, G., Dorochesi, F., Cumsille, A., Valencia, R., Rojas, C., & Seeger, M. (2018). Living at the frontiers of life: Extremophiles in Chile and their potential for bioremediation. *Frontiers in Microbiology, 9,* 2309. https://www.frontiersin.org/articles/10.3389/fmicb.2018.02309/full

Pasteur, L. (1969). *Correspondence of Pasteur and Thuillier concerning anthrax and swine fever vaccinations.* University of Alabama Press. 240 p. ISBN 0817350705.

Peters, T. H. (1995). Revegetation of the copper cliff tailings area. In J. M. Gunn (Ed.), *Restoration and recovery of an industrial region* (Springer series on environmental management). Springer. https://doi.org/10.1007/978-1-4612-2520-1_9

Roy, V., Ruel, J.-C., & Plamondon, A. P. (2000). Establishment, growth, and survival of natural regeneration after clearcutting and drainage on forested wetlands. *Forest Ecology and Management, 129*(1–3), 253–267. https://doi.org/10.1016/S0378-1127(99)00170-X

US EPA. (2016). *Wetlands restoration definitions and distinctions, what is wetland restoration?* https://www.epa.gov/wetlands/wetlands-restoration-definitions-and-distinctions

Walpole, M., & Davidson, N. (2018). Stop draining the swamp: it's time to tackle wetland loss. *Oryx, 52*(4), 595–596.

Chapter 6
Ecological Engineering Tools in Extreme Ecosystems

Margarete Kalin-Seidenfaden (iD)

Abstract Mine waste rock piles and tailings are extreme, oxidative environments. Once the overburden has been scraped off, the reducing sediments or soils have been removed. This leaves newly exposed rock to weathering. The organic sediments need to be replaced. With time, newly created sediments will balance the overwhelmingly oxidative conditions. In ditches and ponds, organic sediments interface between the water body and the exposed rock. Adding organics to ditches and ponds fosters reducing conditions in newly-created sediments. Microbes that flourish under reducing conditions need to be nourished. These anaerobic microbes transform metals in the effluent to authogenic ores. This process has been labeled ARUM or Acid Reduction Using Microbiology. A continuous supply of organics is needed to keep the process going. This is provided by periphyton, phytoplankton, and/or floating, living islands of reeds or cattails. Living islands also protect reducing sediments from wind mixing. This approach has been tested in different acid drainages around the world. A form or ARUM was also used to treat a groundwater plume by dripping urea and sugar into the plume. The urea is hydrolyzed to ammonia which raises the pH with metal and iron precipitation.

Keywords Redox reactions · Reducing sediments · Oxidative environments · Acid Reduction Using Microbiology · Sulfate reduction · In situ treatment · Groundwater · Ureolytic microbes · Heterotrophs

Two types of extreme ecosystems (i.e., terrestrial and aquatic) exist in a mine waste management area, drainages, open water bodies, and terrestrial systems with waste rock and tailings. Drainages are basically artificial streams, brooks, or rivulets. Pit lakes are artificial, as they are not placed into a drainage basin, have no sediments

M. Kalin-Seidenfaden (✉)
Boojum Research Ltd., Toronto, ON, Canada
e-mail: margarete.kalin@utoronto.ca

© The Author(s), under exclusive license to Springer Nature Switzerland AG 2022
M. Kalin-Seidenfaden, W. N. Wheeler (eds.), *Mine Wastes and Water,
Ecological Engineering and Metals Extraction*,
https://doi.org/10.1007/978-3-030-84651-0_6

and no littoral and limnetic zones. A pit lake has no surface water inlets or outlets and is often a hole in the groundwater table.

Every mine has site-specific environmental issues. Indeed, each mine has a different hydrology, geology, and mineralogy. Mines are in different climates, have different physical layouts, and differ with respect to their surrounding environments. The waste- and water- management areas of all these mine sites foster extreme ecosystems, whether the rock is ground or broken, or generating acidic or alkaline drainages. However, the processes for releasing contaminants are the same. For a discussion of weathering and microbial corrosion see Chap. 1.

Boojum Research (Boojum) has developed several ecological tools to counter the forcing functions of these extreme environments. A tool can vary from mine waste management area be it physical (providing structures, adding nutrient with the rain or integration a living supply of organics) addition. In this case, the operation is the minimization or alleviation of forcing functions in mining extreme environments. Tools have been developed to overcome the challenges brought about by oxidative processes occurring at the mineral surface and resulting in acidic or alkaline mine waste effluents.

6.1   Defining the Characteristics of Ecosystems

Boojum has formulated a systematic, multidisciplinary approach to address mine waste challenges. Each project's common features are divided into three phases. Boojum has had the support of mining companies on several pre-feasibility studies (also called scoping studies) which sometimes progressed to pilot projects. They have also had corporate support for several large demonstration programs, some even carried to full-scale.

Phase I (the feasibility study) consists of a literature review on the elements of concern, addressing both their inorganic chemistry and biogeochemistry. A great resource to achieve this is Reitner and Thiel (2011). The literature review is followed by site visits to identify colonizing biota in the effluent streams, on the tailings, and the waste rock piles, starting with the oldest waste, exposed on site since onset of the mine operation. A survey of the undisturbed surroundings is included because resources (seeds, organics, and potential colonizing plants) might be available for site recovery or colonization. The mine history, mineral extraction processes, physical layout, hydrology, and former drainage basins are identified. The physical layout is important as it includes areas where drainages emerge at the foot of the waste rock and tailings piles. In most cases, aerial photographs exist which indicate the original surface drainage, now covered by waste rock piles or tailings. Meteorological data from nearby weather stations, spanning several years, are requested to determine the amounts of precipitation that will fall onto the wastes when the mine operation ceases. In operating mines, it is advisable to produce a water balance of the mill, depending on the sources of freshwater used. Most of this

Table 6.1 The ecological engineering approach to assessing a mine waste management area

Phase I	Site history, mine waste management area
Site Characterization	Physical layout
	Climate, hydrology, surface water and groundwater quality
	Contaminant loadings
	Waste mineralogy, geochemistry contaminant paths and fate
	Contaminant paths and fate
	Process selection
	Existing terrestrial and aquatic ecosystems
	Biological system selection
Phase II	Geochemical and biological reaction rates
Field Testing of Selected Strategy	Define site-specific design criteria for treatment strategies
	Assess feasibility of strategies
	Decision to proceed to Phase III or address missing data
Phase III	Full-scale design, construction, and monitoring of treatment approach
Scale-up of Treatment System	Modify, if necessary, fine-tune system

information is available from the company, found in historic files, or accessible from official government agency websites. (Table 6.1).

Water, tailings, and biological samples found growing on tailings or in drainages are analyzed for contaminant concentrations. The local indigenous flora is identified both on the wastes and in the adjacent undisturbed environment. Growth rates for the most promising organisms are researched or must be determined with laboratory experimentation. The ecological potential to remove the elements of concern (contaminant-removal capacity) is derived from bio-concentration measurements supported by chemical analyses and biological assessments (e.g., growth curves, biomass). These assessments determine the suitability to utilize the ecological processes for the assessed site.

Upon approval of the feasibility report, Phase II (field and laboratory testing) commences. This consists of a combination of laboratory and field testing of materials that trigger precipitation of the contaminants with additions of target substances (mostly organics, fertilizers). Field pilots are carried out on small waste streams to determine residence times and growth rates. Field pilots are generally run for 2 to 3 years and are regularly monitored. Often, adjustments are needed with a one-time addition of nutrients to kick-start the system. Organic materials for the creation of sediments are assessed for availability. For example, in Brazil sugarcane leaves and stalks were used, in Saskatchewan it was large alfalfa bales to create islands in a shallow lake, and in China, rice straw was used. Contaminant loadings are calculated for the whole site, including the larger drainage basin(s) surrounding the wastes, as, at decommissioning, these data may provide evidence of clean, surface run-off. Materials are tested for use in creating additional surface area for the attachment and growth of submerged biota. Time-relevant data were obtained by asking mine site staff to monitor experiments, or, if unavailable, experiments were carried out in the Boojum laboratory, as generally only two trips a year could be financed. Forcing functions were determined in these experiments, and the most cost-effective means to alleviate them is determined, along with the most reasonable monitoring frequency.

A report is then issued to the company and the government regulators, and, depending on their response, the project proceeds to Phase III, with a gradual scale-up, based on the field-pilot results. A monitoring program is carried out for several years, until the ecosystem is homeostatic or has reached self-design.

6.2   Tool for Acid Reduction Using Microbiology

Many publications are available on microbial activity in sediments. Here Stumm & Morgan (1996) microbially-mediated oxidation reactions are used as a reference (pages 473–477). Each microbial reduction reaction proceeds with the oxidation of a compound. Hence, these processes are redox reactions. Each redox reaction in the presence of organic carbon is carried out by different microbes starting with denitrification, followed by iron reduction and sulfate reduction, at which point nearly all labile organics are consumed. If enough oxidized iron is present and the E_h is low enough, the precipitation of authigenic iron minerals will occur. Should this not be the case, sulfate-reducing microorganisms will produce hydrogen sulfide which escapes the sediment as gas – bog gas.

For example, in northern Queensland, Australia, alkaline waste rock drainage was collected in an evaporation pond during the monsoon rains. The company's goal was to remove or reduce the excessive sulfate concentrations in the water. The investigation revealed pHs and E_hs indicative of reducing conditions above the pond sediments. If little or no iron was present in the sediments, sulfate-reducing microbes would convert the sulfate to hydrogen sulfide, leading to bog gas formation. Microbial sulfate reduction might be utilized to reduce the sulfate concentrations in the mine water.

Boojum recommended broadcasting organic matter over the dried evaporation pond bed, which would be filled with alkaline drainage when the monsoons came. As the organic matter decomposed, the sulfates should be released by microbial metabolism. In the absence of iron, hydrogen sulfide gas should be liberated. Boojum failed to point out that without iron, bog gas would be released. Taking the recommendation literally, straw was added to the evaporation pond (Fig. 6.1). The monsoon rains came, flooding the pond. Soon, the intensive smell of hydrogen sulfide emerged, resulting in the sulfate concentrations in the pond water being the lowest ever measured in the discharge. Unfortunately, no data were provided to Boojum. This simple feasibility study illustrates microbial reactions can be used to regulate mine water chemistry.

Microbial sulfate reduction under anaerobic conditions has been shown to remove sulfuric acid and metals from the water column, relegating the mineralized metals to the sediments (Tuttle et al., 1969; Kosolapov et al. 2004). To utilize this

Fig. 6.1 Evaporation pond in northern Queensland, Australia, with straw bales being distributed prior to the monsoon season. (Photograph supplied by Century Mine)

process, anaerobic conditions suitable for sulfate reduction need to be created. Organic materials play a central role in the wetland system, i.e., to chelate metal ions, remove sulfate, increase pH, provide a growth media for microbes (especially sulfate-reducing microbes). The construction of an organic sediment starts with providing straw or hay, not in bales, but distributed loosely to provide the maximum surface area, while allowing some flow through the material. It is important that the added organics contain some easily degradable material (labile component) and provide structure to the sediment (recalcitrant component). Overall, organic materials determine the usefulness of the wetland system to neutralize AMD passively and sustainably.

Although easily degradable organic carbon is desirable for microbial growth (Smith & Kalin, 1991), all materials tested contained both types - recalcitrant and labile compounds, but in different proportions. While the easily-degradable carbon is food for microbes, the added carbon also needs to provide structure, without compacting, so that the drainage can move through it. Alfalfa hay provided the best source for this purpose. Alfalfa had another bonus; it was relatively high in nitrogenous compounds, which are degraded into ammonia by ureolytic microbes. In a similar study by Brown (2002), hay and silage in a 1:1 (weight) mixture, performed best, along with barley and orchard grass silage at reducing sulfate concentrations. Organic material that degrades easily is sometimes needed to 'jump start' the microbial degradation process. High nitrogen content organics such as alfalfa, guinea pig or horse food pellets worked best for this purpose. The compositions are given as a percentage range for each organic compound are given below:

- labile: 1% to 9% lipids, 22% to 40% sugars, starch, and amino acids
- recalcitrant: 17% to 44% cutins, and lignin, and silica, etc. (Smith & Kalin 1991)

To develop this tool, Boojum selected two locations, Elliot Lake and Sudbury, Ontario. In Elliot Lake, the ability of 'artificial' sediments to form secondary minerals was tested. In Sudbury, sediment construction and living, floating islands in tailings acid mine drainage ditches through carbon addition were investigated.

In general, drainages emerge from tailings in a reduced state. Dissolved iron is usually in the form of iron (Fe^{2+}). Waste rock pile drainages generally produce effluents with oxidized iron (Fe^{3+}). Visually, Fe^{2+} oxidation to Fe^{3+} is fast, producing a solid iron hydroxide precipitate along with the co-precipitation of metals (Liu & Kalin, 1994). However, only a fraction of the iron precipitates; a portion stays suspended, passing through a 0.45µm filter. In some drainages, the iron does not readily precipitate, especially in coal waste drainages. Iron can be complexed with organic compounds, e.g., humic acids and other organic molecules. Boojum measured rates of iron oxidation in the field and in the laboratory (Liu & Kalin, 1994). The oxidation of Fe^{2+} to Fe^{3+} is the main driver of pH reduction. Liu and Kalin (1994) used drainages ranging in pH 1.8 to 3.6 and iron concentrations of 10 to 4000 mg. L^{-1} to determine the oxidation rate constant. The rate constant for a particular effluent is needed to determine the residence time needed to reasonably oxidize ferrous iron before treatment. The rate constants calculated for the oxidation of Fe^{2+} varied from 0.00017 to 0.00030 per minute. This suggests that large oxidation ponds would be needed to convert the ferrous to ferric iron. It is recommended that the precipitation behavior of iron be tested in the laboratory for every effluent before oxidation ponds are constructed.

6.2.1   Swampy Drainages and Sediment Mineralization

At the Elliot Lake site is a very slow-flowing acid drainage which forms a shallow pool at the foot of a tailings dam. Boojum selected this site to test if this sequence of oxidation reactions (*sensu* Stumm and Morgan) could be reproduced. Plastic 55-gallon drums, with open bottoms, were filled with loose straw, and inserted into the eroded tailings (Fig. 6.2a). Standpipes were added to the drums to sample porewater for analysis. To be sure that sufficient iron would be present, a handful rusty nails were added.

Monitoring included E_h, pH, and elemental composition of the porewater in the straw/iron/acid drums with drainage. The water from the standpipes was sampled after 3 years and stored under a nitrogen atmosphere to prevent changes in redox. This was necessary as redox conditions change rapidly when groundwater or sediment porewater is exposed to air. Porewater chemistry, pH, and E_h were used as inputs for geochemical modelling with PHREEQC (Parkhurst & Appelo, 2013).

Fig. 6.2 (**a**) Slow-flowing acid drainage forms a shallow pool at the foot of a tailings dam in Elliot Lake. The site was selected to test if secondary minerals could be detected in constructed sediments placed into highly acid water. The drums contained organics and rusty nails with straw. Porewater was sampled under nitrogen gas to examine if it authogenic minerals could be found under the microscope, nettings without rusty nails formed as a control. (**b**) The alfalfa bails were placed in a honeycomb structure to allow reducing and oxidizing heterotrophic microbes to invade the bails. In March 1991 the bails were placed on the ice as access was otherwise not possible. in July 1992 the structures had remained but by August 1993 vegetation had invaded and the bottom pH had improved from 2.3 previous to 5.2. Note the high conductivity which had not changed. (Photographs by Boojum Research)

The pore water from the drums revealed saturation indexes using PHREEQC to determine Saturation Indexes (+SI), then conditions are appropriate for precipitation of minerals, if negative, conditions do not warrant precipitation, and ions remain in solution. Basically, supersaturation needs to be achieved with either a low or a negative redox and/or with a higher pH. Some minerals require both a low redox and high pH.

The analysis of drum porewater, collected without contact with nitrogen indicated positive saturation indices for compounds of iron and sulfur, among others, suggesting that authigenic pyrite was being formed.

The objective, documenting the authigenic mineralization potential of ARUM, had been achieved. The processes described by Stumm and Morgan could be reproduced in the field. However, what Boojum had failed to do was to clearly explain the objectives to the funders of the MEND project, as they expected a green wetland. Nothing green was noted, and funding was cut. Two years later, when Boojum staff returned to the site, the experiment had been destroyed. Only a photographic record of the black, authigenic minerals in the straw remained (Fig. 6.3).

Fig. 6.3 The Elliot Lake drainage pool after mine operators demolished the site and experiments. The black-stained pile close to the road was the only visual evidence of authigenic pyrite. (Photograph by Boojum Research)

6.2.2 Ditches/Open Water

At active mine sites, ditches are constructed to lead drainage from the waste piles toward a chemical treatment plant. The sludge from the neutralization plant is generally pumped back onto the tailings, essentially recycling the problem. These ditches, because they are constructed, have no organic sediments. The MEND project was therefore designed to treat the seepage before it reached the neutralization plant. If the treatment was successful, the water could be discharged with a neutral pH and free of many metals reducing the amount of lime needed for neutralization and thereby also smaller sludge volume.

ARUM testing at the Sudbury site was focused on designing an experimental system to (a) remove excess iron and metals; (b) provide an organic sediment for heterotrophic microbes to establish a reducing environment and (c) provide a continuous source of carbon as food supply for those microbes, which would consume oxygen to produce reducing conditions (Fig. 6.4; Kalin, 1990; Kalin et al., 1991; Fyson et al., 1993).

The tailings seepage emerged with a neutral pH, but quickly oxidized, producing iron hydroxide. The first set of ponds was designed to oxidize the iron from Fe^{2+} to Fe^{3+}, precipitating iron hydroxide and acidifying the seepage. The small ponds after

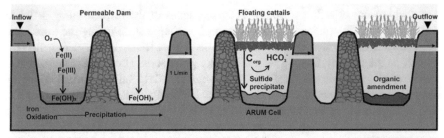

Fig. 6.4 ARUM system development. (**a**) Schematic of ARUM pilot test system for tailings drainage treatment. Seepage water travels from left to right. Oxidation/precipitation ponds are shown in orange, the photograph is showing only a portion of the pond. Curtains are installed to enhance iron settling. In the second pond(photo with floating frame is a settling pond for fine particles of iron hydroxide) ARUM cells in blue in the schematic and in the photo the floating island with vegetation. (**b**) Photograph of the site in Ontario. (Photograph by Boojum Research).

the oxidation cells and the initial ARUM cell were constructed with permeable walls, in case nutrients had to be added, but this was not needed throughout the life of the test cells (Fig. 6.4). The last set of ponds (2) were ARUM cells (Fig. 6.4). The experimental system provided flow control to determine the residence time needed for ARUM to increase the pH and precipitate contaminants. Once an organic sediment is constructed in the ARUM cells, a cover is needed. Floating, living covers provide the water column and sediments with an additional, long-term source of carbon through root exudates and litter. The covers also reduce wind-driven wave action, limiting oxygen penetration into the water column.

After the ARUM cell the next treatment stage is biological polishing. Here, algae, moss and fungi (i.e., periphyton), are grown on substrates to sequester the remaining contaminants. Variations of this design were used to treat the effluents of an abandoned gold mine in Brazil, discussed in Sect. 6.4.1.

6.2.3   Floating Islands: Organic Carbon Supply Supports Reducing Conditions

A tailings pond at decommissioning can be compared to a kettle lake (Fig. 6.5, right side) which reflects the water level of the groundwater table. Often dead tree trunks are sticking out of tailings, especially at the edges of the former valley or lake. Inspecting the tree trunks, one finds a halo of oxidation surrounding the trunk, the beginning of oxidation and acid generation. In natural lakes, organic sediments are found, augmented by floating vegetation along the shore (Fig. 6.5a, photograph). Semi-aquatic vegetation naturally forms floating islands (Mallison et al., 2001). The natural, floating vegetation in bogs and lakes is initially anchored to the shore, but can eventually cover the entire, open water body.

The process of 'landing in' or terrestrialization of water bodies can be observed, exemplified by the schematic in Fig. 6.5b, which describes how ARUM might be used in a decommissioning scenario. First a layer of organic sediments needs to be created at the bottom of the tailings depression. Trees in the valley (depression) should be cut down to prevent creating a 'pipe' through the tailings for oxidation. The shrub, branches and leaves should be left to augment the organic sediments. Tailings are then deposited on the sediment. Once the tailings deposit is completed, a second organic sediment is created, and the shorelines are equipped with floating living islands. Examples of the terrestrialization process are shown in photographs from northern Canada. Figure 6.5a photograph depicts 'landing in.' Figure 6.5b (right side) is a photograph of a seepage from a waste rock pile draining into a depression where a sediment is gradually forming, invaded by semi aquatic vegetation. As the lake/pond ages, vegetation will extend inward from the edges. Upon closure of the mine, the lake/pond will be covered with vegetation, providing reducing conditions for the sediments below. Sharma et al. (2021) describe different functions for floating vegetation islands, some of which might be useful at mine sites.

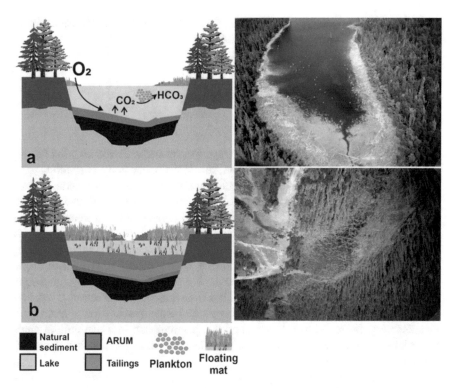

Fig. 6.5 (**a**) Natural 'landing in' lake with schematic and photo showing the 'landing in' or terrrestrialization of a lake/pond seen in the north of Ontario often in the muskeg country. (**b**) the schematic shows the layers, with or without a natural sediment a constructed ARUM sediment should be installed before the tailings are placed in the water body with the same functions as for the ditches. A living, floating cover provides organic carbon for ARUM sediments On the right side of the schematic the brown mass is the drainage from a waste rock pile directed in a small valley with beginning terrestrialisation. (Photographs by Boojum Research)

To select the plants for living, floating covers, the primary focus is tolerance to mine water conditions, either alkaline or acidic. The second focus is to select a plant with a well-developed, loosely-bound root system to collect particulates and produce root exudates. Both *Typha* sp. (cattail) and *Phragmites* sp. (reed grass) tolerate the adverse chemical environments of mine waste water, but *Typha* is better-suited, with a more loosely-bound root system. Initially, the islands need support to float, but with time (3 to 5 years) they become self-supporting, due to the gas produced from the decomposition of the litter (Fig. 6.4b). For more information about cattail islands for mine drainage see Kalin & Smith (1992). Islands can be constructed of other species, such as *Chrysopogon zizanioides* for similar purposes (Kiiskila et al., 2019). Pavlineri et al. (2017) have quantified the effects of using floating, living covers in other remediation situations. Today, floating, living islands are available commercially for a wide variety of applications.

6.2.4 Underwater Meadows: Protecting Contaminated Sediments

In larger water bodies, water turns over (mixes) seasonally, driven by wind mixing and a temperature gradient in spring and fall. If the water body is shallow, wind induced mixing can even re-suspend sediments. In the case of iron-contaminated the sediments in holding ponds or lakes, oxygen can reach the sediment-water interface during periods of mixing. Consequently iron, which had been reduced to ferrous iron in anaerobic sediments, is re-oxidized to ferric iron, leading to a drop in pH. The oxidized iron then re-settles during the summer and fall, again becoming reduced, continuing the cycle. The continuous cycling of iron in lakes with excessive iron in the upper part of the sediments, has been documented by Sulzberger et al. (1990) and Stumm and Sulzberger (1992). In these lakes the seasonal turnover produces a gradual decrease in pH, as oxygenated water reaches the sediment surface annually. Boojum have documented this phenomenon in a one million cubic meter lake as a gradual decrease in pH over the year, with a pH rise during the ice-covered time (Fig. 6.6).

With an ice cover, the iron-hydroxide in the water column settled out onto the sediment. Reducing conditions in the sediment gradually stabilized the iron, raising the pH. As the ice melted, the lake turned over, re-suspending the iron, and re-oxidizing the reduced iron in the top layer of sediments. This was particularly

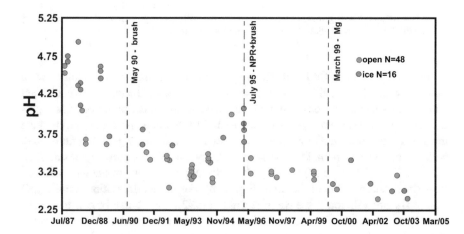

Fig. 6.6 The gradual acidification of a mining lake in northern Ontario, Canada. Throughout the acidification process pHs differed in winter when the late has an ice cover (orange) and summer after turnover of the water body (blue). The dotted red lines are the ecological tools implemented in the lake. Although the data set is small it shows the dramatic changes in the low pH range between 1987 and 2005

prominent in the 1980s before remediation measures were initiated. If the pH decreases much below 3 in acidifying lakes, the oxidized iron remains in solution, gradually producing the distinctive dark red brown color, essentially becoming a leaching solution.

To stabilize contaminated sediments and support a gradual pH increase in the lake, the seasonal cycling of iron has to be prevented. This can be achieved by establishing an acid-tolerant underwater moss meadow over the sediments. The moss, *Drepanocladus sp.*, was found to be an ideal candidate for the job. It is widely distributed in temperate and arctic parts of the world (Wynne, 1944), and grows well in acidic lakes with iron-rich sediments (Satake 2000; Fig. 6.7). *Drepanocladus* was transplanted from a seepage on site to the lake using netting, fastened by an anchor to the sediment (Kalin & Buggeln, 1986). As iron precipitates out of the water column, the moss becomes encrusted, retaining the iron and preventing release to the water column. As the moss dies and decays, the iron is transferred to the sediment. A moss cover also prevents sediments from becoming re-suspended during periods of mixing.

In alkaline ponds and lakes, underwater meadows of charophytes can protect sediments from the re-suspension of the precipitated iron and metals. These algae form dense underwater meadows, growing on the sediment surface (Fig. 6.7a). These fast-growing macrophytes are attached to the bottom of the ponds with root-like rhizoid structures holding the plants in place in sediments. As discussed later (Chap. 7), charophytes promote the precipitation of calcium carbonate with co-precipitated metals. Upon death, the biomass with entrustments sinks into the sediment, providing organic matter for oxygen-consuming microbes. The same function can be shown for other benthic algal and moss species (Fig. 6.7).

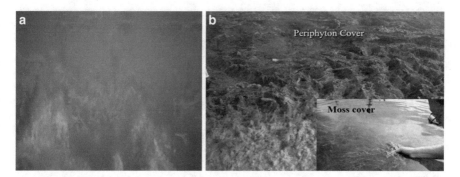

Fig. 6.7 Submerged vegetation like mosses and algae can stabilize ARUM sediments. (**a**) *Nitella fexilis*, a characean alga growing in alkaline water, can cover sediments and accumulate ^{226}Ra and uranium emerging from groundwater through the sediments. These algae can become covered with epiphytes (periphyton) as shown here. (**b**) The green algae (periphyton) growing over acidic pond sediments can also serve this function. *Drepanocladus sp.* (in insert) is an acid-tolerant moss. Photograph insert shows the moss and reduced iron as a brown cloud released when sediments are disturbed. (Photographs by Boojum Research).

6.3 Summary of ARUM Tests in Different Mine Drainages

The water quality changes brought about by ARUM are summarized in Table 6.2. The results are given as removal rates for the metal of concern for different operations and locations. ARUM flow through systems were built and monitored in central Ontario, Nova Scotia, Newfoundland, and northern Saskatchewan. In all locations, the flow and volume of the test cells were measured. A very rough estimate of removal rates in static reactors (2 L jars) at room temperature was determined (Table 6.2 top) and compared to rates calculated from field experimental settings with added straw and low flow rates (1.5 L^{-s} or less; Table 6.2 bottom). The pH of the field systems (2.5–6) varied over the 2–3 years of measurements. Boojum's results show that acidity removal in static reactors ranged between 2.5 and 49 $g.m^{-3}.d^{-1}$, while sulfate removal ranged from 7.1 to 51 $g.m^{-3}.d^{-1}$.

In field flow-through systems, acidity removal ranged from 0.4 to 214 $g.m^{-3}.d^{-1}$, while sulfate removal ranged from 1.2 to 435 $g.m^{-3}.d^{-1}$. Surprisingly, high sulfate removal rates from a waste rock pile in central Newfoundland, Canada were due to additions of straw to an iron-encrusted bog peat. Again, with more iron, more sulfates can be removed as iron-sulfides. These results clearly support the contention that organic carbon sediments alleviate two of the primary forcing functions, low pH, and acidity.

Table 6.2 A summary of water quality achieved in static reactors (2 L jars) and the field pilot systems

AMD Type	Acidity Removal $g.m^{-3}.d^{-1}$	Iron Removal $g.m^{-3}.d^{-1}$	Sulfate Removal $g.m^{-3}.d^{-1}$	Metals Removal $g.m^{-3}.d^{-1}$
STATIC REACTORS				
Ni\Cu	2.5	0.78	8.6	Ni: 0.17
Ni\Cu	5.3	2.5	7.1	(a)
Cu\Zn	8	3.1	11	Zn: 1.2
Cu\Zn\Pb	49	39	51	Zn: 3.8
Cu\Zn\Pb	11	5.5	18	Zn: 0.8
Uranium	41.1	0.005	16	Ni: 0.17
				As: 1.8
FIELD SYSTEMS				
Ni\Cu	0.8	0.04	14	Ni: 0.10
Ni\Cu	2.4	0.04	6.3	Ni: 0.25
Coal	2.8	0.3	3.2	Al: 0.12
Zn\Cu	0.37	(a)	1.2	Zn: 0.15
Zn\Cu	214	0.3	435	Zn: 71
Uranium	(a)	0	3	As: 1.1
				Ni: 0.32

(a) no metals present

Several passive constructed wetlands have been used to generate alkalinity in mine effluents. Pat-Espadas et al. (2018) found some anaerobic wetlands could perform at the rate of 3.5 g acidity.$m^{-2}.d^{-1}$. These data are only comparable if it is assumed that 1 m^3 is equivalent to 1 m^2 of flow (cells are only 1 m deep). Boojum's iron removal rates are of the same order of magnitude as those of Pat-Espanadas et al. (2018), suggesting that similar processes may be at work. Another factor controlling acidity and contaminant removal rates is temperature. All of the static reactor and field measurements were made during the ice-free season, with temperatures varying between 5 and 22 °C.

One of the key criticisms of the ecological approach is the belief that biological systems shut down at low temperatures. However, mine drainage does not freeze, due to its generally high electrical conductivity. Also, adits, which drain from the underground, may flow continuously, but with a lower volume. Microbial activity slows with lower environmental temperatures, but because the chemical reactions of sulfide wastes are exothermic, the drop in effluent temperature is mitigated. For example, in Nunavut and Yukon territories, M. Kalin collected drainage in the winter along waste rock piles and tailings dams, which were thought to be frozen. She also observed tailings seeps on Baffin Island which contained algal growth. Exploration drill holes and small springs were also green with algal growth.

6.4   ARUM Applications

6.4.1   Brazilian Mine Discharge

An abandoned gold mine portal discharged acid drainage to the Das Vehlas River, Brazil. Ponds were constructed into the hillside, starting with two precipitation ponds followed by three ARUM ponds (Fig. 6.8; Kalin & Caetano Chaves, 2003). Sugar cane litter was used to create organic sediments in the ARUM ponds. Cattails were planted on Styrofoam™ floats over the newly-created organic sediments. Monitoring data continued for 1 year, demonstrating the success of the system. A personal communication with V. Ciminelli of the University of Minas Gerais later established that the system had been working for 25 years before it was destroyed by the local population (Fig. 6.8).

6.4.2   Dried Out Lake Sediment: Re-solubilization of Elements

Sediments from several dried-out lakes were studied in northern Saskatchewan. The original lakes were drained as the groundwater table was lowered to construct an open pit to reach uranium ore. The lakes received water only from atmospheric precipitation. Organic sediments in the lakes contained metals and authogenic

Fig. 6.8 The ARUM system in Brazil. (**a**) The portal of the abandoned gold mine. (**b**) The precipitation ponds. (**c**) Cattails transplanted to Styrofoam™ floats as living cover for ARUM cells. Photograph shows initial system. Which was later replaced (Kalin & Caetano Chaves, 2003). (Photographs by Boojum Research)

pyrite, the natural result of ARUM, noticeable due to small acidic seeps occurring after rainfall or snowmelt.

At close-out of the mining operation, the rising groundwater would re-fill the lakes, flooding the sediments. Because the buffering capacity of the groundwater was very low, the newly re-filled lakes would turn acidic. The environmental concern was that nickel, also present in the sediments, would be released, and exceed the allowable environmental concentration limit. This would necessitate chemical treatment of the lake water, a costly undertaking. The objective of this work was to address two issues. Which of the lakes contained sediments with nickel-bearing strata, and, if so, could preventative action be taken to retain the nickel within the newly flooded sediments?

The first question was addressed with an extensive lake sediment sampling campaign, which included digging pits to a depth of 1 m and sampling the stratigraphic layers. Stratigraphic and pedological assessments along with chemical analysis identified those lake sediments which needed neutralization and nickel immobilization. The problem was isolated to mainly one lake. Cores were taken, and carefully introduced to plastic columns, with sampling ports in each of the strata. The columns were stored in an industrial cellar at relatively low temperatures. The columns were flooded from the bottom upward. Each of the port effluents was sampled for the presence of nickel, identified through previous chemical analysis and a pedological assessment. This installation was run and monitored for 5 years (Fig. 6.9a).

The second question regarding the prevention of possible nickel release, could be answered by creating ARUM sediments. The dried-out lake sediments contained no organic matter. To test this, lake sediments and easily-degradable organic material

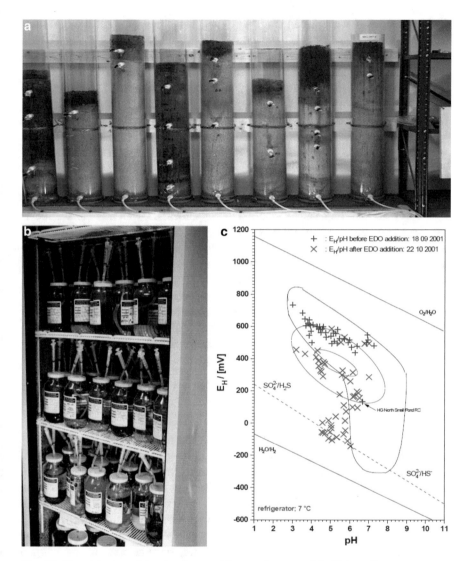

Fig. 6.9 Experiments to determine future acid-generating potential of lake sediments accompanied by nickel release to the rising groundwater table. (**a**) Soil cores were obtained from several dried-out lakes to simulate rising groundwater table. Sampling ports were installed in different core strata to simulate/document potential nickel release. (**b**) The jars in the refrigerator were equipped with mini piezometers in the sediment and connected to syringes to obtain porewater before and after the addition of potato peel waste (easily degradable organics, EDO). (**c**) The results were plotted into the phase diagram to identify which groups of microbes would be active, based on the extensive database of surface and groundwater of Baas Becking et al. (1960). The pH/E_h couples of the porewater before the organics were added are marked in blue and in red after the addition. (Photographs by Boojum Research)

(potato peel waste) were placed into 1 L jars with sampling ports (small tubes) serving as porewater samplers. The potato waste would, hopefully, foster ARUM microbes to precipitate the nickel. Both the porewater in the sediment and the overlying water were monitored for nickel concentration, pH, and E_h (Fig. 6.9b).

The jars were filled with groundwater shipped from northern Saskatchewan to Toronto. In Fig. 6.9c the results are shown in an E_h-pH Baas Becking diagram (Baas Becking et al., 1960). The pH and E_h pairs are plotted against each other, before the addition of potato waste (blue crosses) and after (red xs). More than 90% of all samples to which organic carbon was added showed a rising trend in pH, and a decrease in E_h (Boojum Research, 2006). After one month the pH and E_h pairs moved in the direction which promoted thiobacteria. Sulfate-reducing bacteria, however, need even lower pH (<4) and negative E_hs (i.e., a region not yet reached).

It appeared that the results supported the hypotheses. Given that the sediments had been dried out, the concern was raised by the client that the needed microbes would be dead. Boojum's assurance was insufficient, particularly for the governmental regulators. Later, Boojum received a call from the project leader saying that the lakes were flooded, and the projections were correct. Boojum had located the contaminated lakes correctly and the nickel concentrations were acceptable. However, they had problems with zinc from galvanized pipes needed in the operation. Unfortunately, the project leader had a fatal accident, and the project was terminated. Here, as in Australia, Boojum provided evidence that organics are a key component in initiating a change in the existing mine water conditions.

6.4.3   In Situ ARUM Application

In northern Ontario, a polymetallic mine was developed on a peninsula, between two lakes, one of which was a trophy fishing lake, and the other, a contaminated experimental lake. Boojum was hired to assist with the decommissioning, since regulators and the company could not find a suitable place for a conventional chemical treatment plant or the resultant sludge. Sludge production was estimated between 1200 m^3 and 3,000 m^3 per year with high density sludge production projected to last for 1000 to 35,000 years (Kalin, 2002). At that time, the mine operator and Boojum had absolutely no knowledge of the groundwater conditions. Boojum engaged a hydrologist and a geochemist to address this challenge.

Between 1986 and 2000 more than 110 piezometers were installed in and outside the tailings area. The groundwater plume encountered in some piezometers contained close to 1 $g.L^{-1}$ each of iron and zinc, and about 20 $g.L^{-1}$ of sulfur and was expanding in strata at a depth of 16 m. This deep groundwater plume was moving toward a small, lake to the north of the tailings called Mud Lake. A groundwater model was constructed, based on the stratigraphy and the water level behavior in the tailings.

In 1998, the software package, visual Modflow, was used to construct a numerical model of the larger drainage basin bounded by the trophy lake to the south

Hariharen & Uma Shankar (2017). This three-dimensional groundwater flow model is an industry standard and had been subject to extensive verification and validation studies (Harihan & Shankar, 2017). The model predicted that groundwater was moving in several directions. It was feared that the plume would travel towards the trophy fishing lake, a highly, undesirable outcome. Flows toward the direction of the trophy fishery lake was relatively small at about 3,700 $m^3.a^{-1}$.

The global, site-wide model provided an estimate of the dilution potential from the groundwater system of the various watersheds surrounding the mine and tailings sites. It was determined that there are about 900,000 $m^3.a^{-1}$ of uncontaminated groundwater available for diluting about 150,000 $m^3.a^{-1}$ of groundwater in the mine site and tailings. Most of the flow from the tailings was via the bedrock canyon to Mud Lake to the north, at about 18,000 $m^3.a^{-1}$. The remaining flow paths accounted for 7000 $m^3.a^{-1}$. There was about 9000 $m^3.a^{-1}$ flowing from the tailings pond to Mud Lake.

The results of transport modelling using zinc as the representative contaminant indicated a breakthrough of contamination into Mud Lake in 5 to 10 years, with a predicted zinc concentration of 60 $mg.L^{-1}$ at the inflow of this groundwater into Mud Lake after 20 years (Fig. 6.10). These predictions agree well with field samples taken from piezometers in the flow path. The estimated loading of 1.3 tonnes of zinc per annum into Mud Lake was compared to the 0.5 tonnes per annum previously estimated for discharge into the trophy fishing lake. As the trophy fishing lake is a very large lake at the headwaters of the English River, this plume could be considered minor. The loadings estimated are very conservative, since it was assumed that an average concentration of 200 $mg.L^{-1}$, representing the upper range of zinc concentrations within the tailings, were distributed throughout the tailings (Fig. 6.10).

The modeling demonstrated that the contaminant pathways were south to a diversion ditch and then to the contaminated experimental lake, which was designated as a biological polishing lake. One plume moved to the north via the bedrock canyon to Mud Lake, not toward the trophy fishing lake. After a few years, the groundwater emerged into Mud Lake, where the pH dropped from between pH 5 and 6 prior to the emergence of the plume to pH 2.5 within 3 months. This triggered an interest in treating the groundwater plume to prevent or lessen its effects on Mud Lake.

To treat the groundwater plume, Boojum decided to feed nutrients into the plume to activate and grow ureolytic microbes which would raise pH and precipitate contaminant metals (see equations). These microbes would degrade urea to ammonia in the plume which would raise the pH. Sugar should activate the heterotrophic microbes along with the sulfate reducers.

Laboratory experimentation with this type of drainage is not possible due to its reactivity with oxygen. Therefore, Boojum supported a geochemical modeling study of the expected reactions (Fleury, 1999), including the well-documented reactions of urea-degrading microbes:

$$CO(NH_2)_2 + 2H_2O \Rightarrow 2NH_4^+ + CO_3^{2-}. \tag{6.1}$$

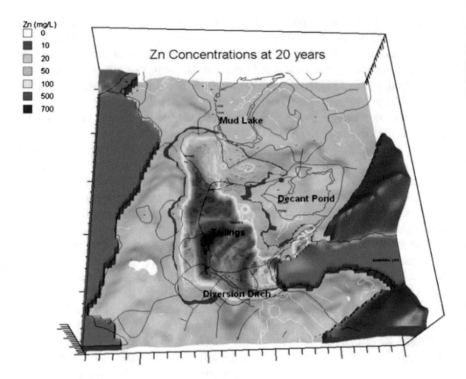

Fig. 6.10 Verification of the model using the projected zinc concentrations. The 20-year projections are taken from the startup of the operation of the mill with 100 mg.L^{-1} at the edge of Mud Lake. (SCIMUS, 2003)

$$CO_3^{2-} + H^+ \Rightarrow CO_3^- \qquad\qquad (6.2)$$

The indigenous microbial flora of the drainage water was quantified and found to be relatively rich in microbial diversity (Lau et al., 2001), although no ureolytic microbes were found. The agricultural literature predicted that ureolytic microbes not active below pH 4.0 (Burton & Prosser 2001), a pH which could be encountered the groundwater plume. Experimentation with groundwater was complex, but tests with a commercial enzyme in the negative redox range needed to be carried out in the field. It was also likely that sulfate-reducing bacteria were present, as predicted (Fleury, 1999).

A pilot test area was found which had a reasonably high groundwater flow rate, to follow the expected pH increase (Fig. 6.11a, b). The tests were scaled up on the shores of Mud Lake into which the plume discharged. At this location, the 16 m deep plume was intercepted, and an injection protocol was implemented. A schematic of the injection system is depicted in Fig. 6.12a. The photographs (Fig. 6.12b, c) show the standpipes used to monitor the injected plume movement on a grid. Piezometers in the muskeg area on the shores of Mud Lake intercepted the plume in

Fig. 6.11 (**a**) Yan Gan is shown performing the field tests on the activity of a commercial ureolytic enzyme. (**b**) Murray Johnson shown installing injection standpipes in the sandy area to a depth of 0.5 m which served as the field pilot. A urea-sugar mix was injected about 15 to 20 cm into the sand. (Photographs by Boojum Research)

a stratum 16 m below the floating muskeg. Below the muskeg a thick layer of gyttja sediment was present into which the microbial food was injected. This made it possible to drip a urea-sugar mixture into the plume without oxygen contact. The installation was in operation by fall 2000, at which time a sampling campaign was carried out to determine all background water quality parameters.

By August 2002, 529 m^3 of plume drainage had been injected with 16.1 m^3 of a urea-sugar mixture. Monitoring ceased in summer 2003, after a final injection of 2 m^3 urea-sugar slurry together with 58 m^3 of drainage. The monitoring data

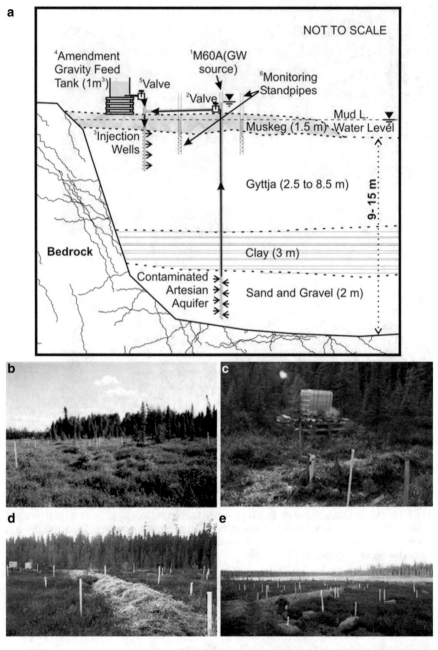

Fig. 6.12 (**a**) Schematic of the *in-situ* ground water drainage treatment pilot test. (**b**) Top left photograph shows the untouched floating muskeg with piezometers. Piezometer M60A, in the foreground, was central to the experiment as it behaved like an artesian well producing acid drainage. (**c**) the top right photograph shows the head feeding tank containing the urea-sugar mixture. (**d**) Bottom right photograph displays the winterization of the T-junction between the injection well and the food supply. (**e**) Bottom right photograph shows winterizing of the T-junction to the well M60A (at the long end of the T) with the contaminated lake in the background. (Photographs by Boojum Research)

indicated that the process was working; organic carbon was consumed, ammonia was produced, and as expected, the metals were precipitated. A summary of this work is provided in Kalin et al. (2008).

References

Baas Becking, L. G. M., Kaplan, I. R., & Moore, D. (1960). Limit of the natural environment in terms of pH and oxidation-reduction potentials. *Journal of Geology, 68*, 243–284. Retrieved from: www.jstor.org/stable/30059218

Boojum Research Ltd. (2006). Key Lake Operations: Soil/Sediment interaction with ground and surface water during lake recovery. Potential biological treatment measures. *Produced for Cameco Corporation.* Retrieved from: https://zone.biblio.laurentian.ca/handle/10219/2910

Brown, A. (2002). *A comparison study of agricultural materials as carbon sources for sulphate reducing bacteria in passive treatment of high sulphate water.* MSc. University of British Columbia.

SCIMUS, A. (1986). A systematic approach to the analysis of waste management systems., *Canadian Nuclear Society, 2nd International Conference on Radioactive Waste Management,* Toronto, p 548–557.

Burton, S. A., & Prosser, J. I. (2001). Autotrophic ammonia oxidation at low pH through urea hydrolysis. *Applied and Environmental Microbiology, 67*(7), 2952–2957.

Fleury, T. L. (1999). *A geochemical modeling study of the effects of urea-degrading bacteria on groundwater contaminated with acid mine drainage.* MSc thesis. Department of Geology, University of Toronto, Toronto, Canada. Retrieved from https://hdl.handle.net/1807/13439

Fyson, A., Kalin, M., and Smith, M. P. (1993). Microbially-mediated metal removal from acid mine drainage, In *FEMS Symposium, Metals-Microorganisms Relationships and Applications,* Metz, France, May 5–7 (11 p). Retrieved from: https://link.springer.com/chapter/10.1007/978-94-017-1435-8_47

Hariharen, V., & Uma Shankar, M. (2017). A review of visual MODFLOW applications in groundwater modelling. *IOP Conference Series: Materials Science and Engineering, 263,* 32025. Retrieved from: doi:https://doi.org/10.1088/1757-899x/263/3/032025

Kalin, M. (1990). Treatment of acidic seepage employing wetland ecology and microbiology. Final Report. pp 171. Retrieved from: https://zone.biblio.laurentian.ca/handle/10219/2984

Kalin, M. (2002). Slow progress in controlling AMD: a perspective and a new approach. In *New Technology Implementation in Metallurgical Processes: as held at the 41st Annual Conference of Metallurgists of CIM(COM 2002),* pp. 323–341.

Kalin, M. (2004). Slow progress in controlling acid mine drainage (AMD): A perspective and a new approach. *Peckiana, 3,* 101–112.

Kalin, M. & Buggeln, R. (1986). Acidophilic aquatic mosses as biological polishing agents. Report for Dr. Ron McCready, CANMET, Energy, Mines and Resources. Retrieved from: https://zone.biblio.laurentian.ca/handle/10219/3051

Kalin, M., & Caetano Chaves, W. L. (2003). Acid reduction using microbiology: Treating AMD effluent emerging from an abandoned mine portal. *Hydrometallurgy, 71*(1–2), 217–255.

Kalin, M., & Smith. (1986). Biological polishing agents for mill waste water. An example: *Chara.* In R. W. Lawrence, R. M. R. Branion, & H. G. Ebner (Eds.), *Fundamental and applied biohydrometallurgy* (p. 491). Elsevier.

Kalin, M., & Smith, M. P. (1992). The development of floating Typha mats. *Proceedings of the IAWPRC Conference on Wetland Systems in Water Pollution Control, Wetlands Downunder,* Sydney Australia, 9 pp.

Kalin, M., Cairns, J., & Wheeler, W. N. (1991). Biological alkalinity generation in acid mine drainage. In *Proceedings of the Second International Symposium on the Biological Processing of Coal*, San Diego, California, May 1-3. (p. P105-P112) Retrieved from: https://inis.iaea.org/search/search.aspx?orig_q=RN:23062656

Kalin, M., Fyson, A., & Smith, M. P. (2008). Metal contaminated ground water from base metal tailings: results from a field pilot-scale in situ treatment test. In J. Wiertz (Ed.), *Water in mining, proceedings of the international congress on water management in the mining industry* (pp. 357–370). GECAMIN Ltd..

Kiiskila, J. D., Sarkar, D., Panja, S., Sahi, S. V., & Datta, R. (2019). Remediation of acid mine drainage-impacted water by vetiver grass (*Chrysopogon zizanioides*): A multiscale long-term study. *Ecological Engineering, 129*(February), 97–108. Retrieved from:. https://doi.org/10.1016/j.ecoleng.2019.01.018

Kosolapov, D. B., Kuschk, P., Vainshtein, M. B., Vatsourina, A. V., Wießner, A., Kästner, M., & Müller, R. A. (2004). Microbial processes of heavy metal removal from carbon-deficient effluents in constructed wetlands. *Engineering in Life Sciences, 4*(5), 403–411. Retrieved from:. https://doi.org/10.1002/elsc.200420048

Lau, P. C. K., Bergeron, H., & Kalin, M. (2001). Bacterial consortia in a ground water plume from acid generating tailings. *Proceedings of the Fourth International Symposium on Waste Processing and Recycling in Mineral and Metallurgical Industries, MET SOC 40th Annual Conference of Metallurgists of CIM* (pp. 371–379).

Liu, J. Y., & Kalin, M. (1994). Study of ferrous oxidation process in AMD seepage. *Proceedings of the International Land Reclamation and Mine Drainage Conference*, Pittsburgh, Pennsylvania, April 24-29. Retrieved from: http://pdf.library.laurentian.ca/medb/Authors/Kalin/Posters/P8.pdf

Mallison, C. T., Stocker, R. K., & Cichra, C. E. (2001). Physical and vegetative characteristics of floating islands. *Journal of Aquatic Plant Management, 39*, 107–111. Retrieved from: https://www.researchgate.net/publication/237555790

Parkhurst, D. L., & Appelo, C. A. J. (2013). Description of input and examples for PHREEQC version 3—A computer program for speciation, batch-reaction, one-dimensional transport, and inverse geochemical calculations. *U.S. Geological Survey Techniques and Methods,* Book 6, U.S. Geological Survey Techniques and Methods.

Pat-Espadas, A. M., Portales, R. L., Amabilis-Sosa, L. E., Gómez, G., & Vidal, G. (2018). Review of constructed wetlands for acid mine drainage treatment. *Water (Switzerland), 10*(11), 1–25. Retrieved from:. https://doi.org/10.3390/w10111685

Pavlineri, N., Skoulikidis, N., & Tsihrintzis, V. A. (2017). Constructed floating wetlands: A review of research, design, operations and managements aspects, and data meta-analysis. *Chemical Engineering Journal, 308*, 1120–1132. Retrieved from: https://www.sciencedirect.com/science/article/abs/pii/S1385894716313857?via%3Dihub

Reitner, J., & Thiel, V. (Eds.). (2011). *Encyclopedia of Geobiology* (Encyclopedia of earth science series). Springer.

Satake, K. (2000). Iron accumulation on the cell wall of the aquatic moss. *Hydrobiologia, 433*, 25–30.

SCIMUS Inc. (2003). A summary of modeling studies of the South Bay Mine site. *Report prepared for Boojum Research Ltd.* NOT IN LIBRARY YET

Sharma, R., Vymazal, J., & Malaviya, P. (2021). Application of floating treatment wetlands for stormwater runoff: A critical review of the recent developments with emphasis on heavy metals and nutrient removal. *Science of The Total Environment, 777*, 146044.

Smith, M. P., & Kalin, M. (1991). Floating *Typha* mat populations as organic carbon sources for microbial treatment of acid mine drainage. *Proceedings of the IX International Symposium Biohydrometallurgy,* Troia, Portugal, p. 454.

Stumm, W., & Morgan, J. (1996). *Aquatic chemistry: Chemical equilibria and rates in natural waters* (3rd ed.), Wiley.

Stumm, W., & Sulzberger, B. (1992). The cycling of iron in natural environments: Considerations based on laboratory studies of heterogeneous redox processes. *Geochimica et Cosmochimica Acta, 56*(Iii), 3233–3257. Retrieved from http://linkinghub.elsevier.com/retrieve/pii/001670379290301X

Sulzberger, B., Schnoor, J. L., Giovanoli, R., Hering, J. G., & Zobrist, J. (1990). Biogeochemistry of iron in an acid lake. *Aquatic Science, 52*, 56–74. Retrieved from:. https://doi.org/10.1007/BF00878241

Tuttle, J. H., Dugan, P. R., & Randles, C. I. (1969). Microbial sulfate reduction and its potential utility as an acid mine water pollution abatement procedure. *Applied Microbiology, 17*(2), 297–302.

Wynne, F. (1944). Studies in *Drepanocladus*. II Phytogeography. *The American Midland Naturalist, 32*(3), 643–668. https://doi.org/10.2307/2421242

Chapter 7
Biological Polishing Tool: Element Removal in the Water Column

William N. Wheeler, Carlos Paulo, Anne Herbst (iD), Hendrik Schubert, Guenther Meinrath, and Margarete Kalin-Seidenfaden (iD)

Abstract Mining effluents are colonized by algae, moss, fungi and higher plants. Many of these can extract and hyperaccumulate metals. This process is known as biological polishing. Algae growing in mine effluents can sequester metals through adsorption onto cell walls and by absorption. Charophytes are a specialized group of algae that grow in temperate freshwaters and in alkaline mine waste water. They can hyperaccumulate radium and uranium, as well as many other cations. Their calcium carbonate infused cell walls are also a sink for carbon dioxide, making them very useful biological polishers. Boojum Research has studied mine waste water in many places and noted a number of situations where algae and charophytes may be useful for cleansing metals and radionuclides.

Algae are capable of surviving and growing in effluents with pHs as low as 0.8, but due to the lack of dissolved inorganic carbon (DIC), do not grow well. To give them more DIC, the pH needs to be above 4.0, where bicarbonate becomes the dominant form. To accomplish this, the pH must be elevated. This can be accomplished by a number of means, but the most ecologically sensitive method is to dissolve magnesium metal or alloys. This technology is discussed in detail.

W. N. Wheeler · M. Kalin-Seidenfaden (✉)
Boojum Research Ltd., Toronto, ON, Canada
e-mail: margarete.kalin@utoronto.ca

C. Paulo
School of the Environment, Trent University, Peterborough, ON, Canada

A. Herbst
Department Maritime Systeme, Interdisziplinäre Fakultät, Rostock, Germany
e-mail: anne.herbst@uni-rostock.de

H. Schubert
Universität Rostock, Biowissenschaften Lehrstuhl für Ökologie, Rostock, Germany
e-mail: hendrik.schubert@uni-rostock.de

G. Meinrath
Head, RER Consultants Passau, Passau, Bavaria, Germany

© The Author(s), under exclusive license to Springer Nature Switzerland AG 2022
M. Kalin-Seidenfaden, W. N. Wheeler (eds.), *Mine Wastes and Water, Ecological Engineering and Metals Extraction*,
https://doi.org/10.1007/978-3-030-84651-0_7

Keywords Biological polishing · Charophytes · Algae · Periphyton · Extracellular polysaccharides · Algal blooms · Carbon sequestration · Radium · Magnesium alloys · Halophytes · Selenium · Tufa · Modelling · Radium

Biological polishing is a term that describes the use of plant and microbial biomass to filter contaminants from waste water. These self-maintaining absorbants are tolerant of extreme environments and are integrated into complete biological treatment sytems. These living organisms whereby they extrude or surround themselves with adsorbents, many benefits to the ecological engineer. 1. Sequestration of ionically charged contaminants on oppositely charged cell walls. 2. Sequestration and flocculation of particulates with extracellular polysaccharides. 3. Provision of organic carbon for ARUM (Acid Reduction using Microbiology) sediments (with associated contaminants). 4. Carbon sequestration and removal from the atmosphere. 5. Photosynthesis and aquatic pH buffering. Each of these benefits will be touched on as it applies to mine waste water, both alkaline and acidic.

Collectively, aquatic vegetation consists of single-celled algae (phytoplankton), multi-celled algae, moss and fungi (periphyton), and submergent, emergent or floating, photosynthesizing higher plants. Planktonic algae come in two size classes – those greater than 2 μm (phytoplankton), and those between 0.2 and 2 μm (picoplankton). Multicellular periphyton can be mm to meters in size. While periphyton can be observed with the naked eye, phytoplankton and picoplankton are more microscopic, yet can contribute over 70% of the primary productivity in freshwater lakes (Stockner & Antia, 1986; Stockner, 1988).

A survey of several acidic pit lakes in the former coal mining area of Lausitz Germany (Steinberg et al., 1996) found picoplankton of varied composition (chrysophytes, chlorophytes, diatoms, and cyanobacteria). They also found that picoplankton were present only if the pH was above 4.0–4.5 (Steinberg et al., 1998). Attempts have been made to find picoplankton in lakes with pHs below 4.0 by Boojum Research (Boojum), but the results have been inconsistent. In lower pH ranges iron precipitate result in fault counts. This is evidenced by Baker et al. (2004) who found fungi and active eukaryotes in acid mine drainage with pHs as low as 0.8, in warm water of 30–50 °C, and rich in metal ions. They, (Baker et al. 2004), used several methods to detect biota, most of them of a genomic nature.

Since algae seem to be ubiquitous in aquatic environments, from extremely alkaline to very acidic, the question arises, how can these plants be used to trap and remove contaminants in quantities great enough to be significant? The role of the ecological engineer is to provide the chemical and physical conditions necessary to remove the forcing functions limiting their growth and production. This means providing them with inorganic carbon and other nutrients, suitable substrates, and enough light for photosynthesis.

One of the ways algae can polish waste water is by acting as flocculants to adsorb contaminants and contaminated particulates onto living surfaces and exudates. As the flocculated material grows in size, it becomes denser and sinks, carrying particulates and contaminants with it. According to Stumm and Morgan (1996, p. 818), particles (organic or inorganic) in freshwater systems play:

> a commanding role in regulating the concentrations of most reactive elements and of any pollutants in soils and natural water systems and in the coupling of various hydrochemical cycles.

Conventionally, flocculating agents (aluminum and iron salts or polyelectrolytes) are added to waste water to settle particulates out of the water column. However, algae, fungi and microbes can perform a similar function. They, and their exudates, extracellular polysaccharides (EPS), are electrically-charged anions, attracting ions, precipitates, and other charged molecules. As the particles aggregate, they become heavier, sinking to the sediment (Buffle & van Leeuwen, 1992).

All algae and microbes excrete large organic compounds called extracellular polysaccharides. These molecules form colloids, gels or (emulsion-like particles) which aggregate around the cells thereby increasing surface area and volume. The chelating properties of EPS are described for *Chlorella* spp. by Kaplan et al. (1987). The cells, along with the EPS, tend to aggregate (flocculate), until a critical mass is reached and the cells, EPS and attached particulates and ions, are large enough to sink to the lake or ocean bottom or to deeper layers in the water column. EPS can also become separated from phytoplankton cells, and aggregate particles and compounds without the original algal or microbial cells (Casado-Martinex, 2013).

7.1 Which Elements Can Be Biologically Polished?

Mine managers often ask the question – can algae clean every element? Algae 'clean' metals and elements by adsorbing them onto their cell walls, and/or absorbing them into their cells. Living cells have reactive cell walls, and can carry surface charges (electric fields) which can be negative or positive. These surface charges attract oppositely-charged ions and charged particles. These cell walls are composed mainly of polysaccharides and carbohydrates, e.g., cellulose, xylan, and mannans (Neihof & Loeb, 1972; Myers et al., 1975; Lobban & Wynne, 1981), with a negative charge on their surfaces. The negatively-charged groups attract cations such as Zn^{2+}, Cu^{2+}, Al^{3+} and U^{4+} (Barker & Banfield, 1998; Marques et al., 1990). The polysaccharide backbone contains many other side groups, ligands such as amino groups, which bind metals (Sterritt & Lester, 1979). This ability to sequester metals makes many of these organisms metal hyper-accumulators (Gale & Wixson, 1979; Mann et al., 1988; Juwarkar et al., 2010).

The question is particularly relevant given the boom in mining for REE (Rare Earth Elements). Many of the rare earth elements are trivalent and have been shown to be absorbed by algae, such as Gadolinium by *Chlamydomonas* (Aharchaou et al.,

2020), and Lanthanum by *Chara* (Li et al., 2008). Breuker et al. (2020) report on microorganism biosorption of REE elements, including a couple of algae. Casado-Martinex (2013) states that REEs in aquatic plants displace calcium and have a high affinity for phosphate groups in biological macromolecules. In another study, Manusadžianas et al. (2020) found that of the 12 REE elements studied, there might be a possible linkage between the chemical/physical properties of REEs and their biological effects. Ionic radius and atomic number seem to correlate with toxicity to a charophyte, *Nitellopsis*.

Crist et al. (1988) approached the biological uptake and adsorption of metals purely from a chemical perspective by titration. They defined two processes by which metals interact with algae, a fast surface reaction (seconds) and a slow, diffusion reaction. The fast reaction was attributed to adsorption onto algal cell walls. The slow reaction was diffusion into the cell. The number of adsorption sites on the cell wall is determined by wall structure and composition. Crist et al. (1988) further noted that for each species of algae, as the pH increased from 4.5 to 5.5, the number of adsorption sites on the cell walls increased, but not by the number of sites freed up by proton removal. They concluded that not all new sites created by proton removal are effective for metal adsorption. For a given alga, the maximum number of sites is relatively constant within a family of metals (e.g., alkali-metals, transition metals), so that metals in each family should have relatively similar bioconcentration factors.

This seems to correlate with data found by Boojum while studying the adsorption/absorption data for the green alga, *Chara* and periphyton in mining waste water and pristine ponds. Boojum had a large data set of characean biomass and summarized the elemental concentrations, grouping them according to the element family. It reflects the bioconcentration possibilities where *Chara* biomass might assist in reducing the concentrations in waste water.

Boojum Research postulated that the fundamental chemical characteristics of each metal or contaminant may play a role in particle formation, adsorption and/or absorption onto/into the biomass. Boojum compared the charophyte biomass samples collected in alkaline mine waste water and clean water, against a set of periphyton precipitate complexes (unwashed periphyton with precipitates) found growing in acidic water during the spring and summer. The biomass of both was dried, acid digested, and analyzed for 25 elements with ICP (Inductively-Coupled Plasma Spectroscopy). Water samples collected next to the algae were filtered (0.45μm), and similarly analyzed. Concentration factors were expected to be greater in the acid-grown biomass than in the alkaline-grown biomass, as pH has a strong influence on cell wall reactivity. However, when examining the graphs (Fig. 7.1a, b), it appears that biological polishing of different metals is essentially chemical in nature, influenced by the biota and their cell wall characteristics. While no statistical significance can be claimed for the data sets consisting of 14 and 24 water and biomass samples, respectively. The histogram patterns show similarities between the alkaline and acid-grown algae when the elemental groups are considered, reflecting their physical and chemical characteristics. It would be an interesting project to address the characteristics of the cell walls supporting the biological

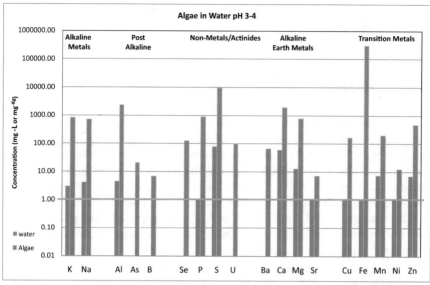

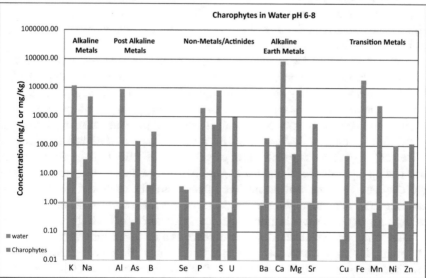

Fig. 7.1 Concentration of major elements in charophytes washed free of periphyton (attached algae) and the water in which they were growing. (**a**) In an acidic lake contaminated by a Cu/Zn mine, 14 samples of periphyton-precipitate complexes and the surrounding water were collected over a 4-month period during the growing season and were analyzed by ICP. (**b**) Charophyte bio-mass samples and surrounding water (n = 24) from 5 contaminated and 3 pristine water sites

polishing process in acid and alkaline water given that pH is an important driver within the adsorption process.

7.1.1 Biological Polishing Model

To understand the biological polishing process and use it to predict how much contaminant could be removed from the water column at a given site, Boojum developed a computer model, in which algal growth characteristics, photosynthesis data, and water chemistry were entered to predict the capacity of the biological polishing system to remove contaminants. With this tool, it was hoped to better understand the bio-geochemical interactions within extreme ecosystems.

The computer model describes the growth of algae (periphyton) and their interaction with mine drainage. Input parameters included: surface areas of curtains for attachment, amounts of fertilizer to be added, and seasonal light and temperature patterns required for optimum growth. As plants photosynthesize (raising pH), and adsorb contaminants, they change the water chemistry. Changes in E_h (oxygen is a byproduct of photosynthesis) and pH of the water as a result of photosynthesis significantly alter the solubility of metals in solution, possibly precipitating metals as hydroxides and carbonates.

The computer model is an attempt to form a bridge between 'top down' and 'bottom up' approaches, combining a mechanistic, theoretical perspective of the key biogeochemical processes operating in mine waste water polishing ponds and the empirical approach for quantifying the complex ecological growth processes. The changes that can be brought about by algal growth in the water were evaluated with PHREEQE (Parkhurst & Appelo, 2013), a model that was designed to perform a wide variety of aqueous geochemical calculations. PHREEQE is based on an ion-association aqueous model and has capabilities for speciation and saturation-index calculations. This mechanistic/empirical model must be calibrated and verified in the field. The first attempt to use the model on a 1 million cubic meter acidic lake with a turnover of 3 years had limited success (Romanin, 1994; Kalin et al., 1995).

The computer model was designed to serve as a management tool for biological polishing, in that it would allow the user to determine the standing crop and time required to 'clean' a particular effluent. The model exists in FORTRAN and it is now at least 25 years old. Progress in climate change modelling has taken great leaps forward in the last couple of decades, which means the model could be drastically improved. Boojum is in discussions with researchers at the Swiss Federal Institute of Aquatic Science and Technology to see if it can be developed further (Fig. 7.2).

Fig. 7.2 Flow chart of the computer program to combine a mechanistic, theoretical perspective of the key biogeochemical processes operating in mine waste water polishing ponds and an empirical approach for quantifying the complex ecological growth processes

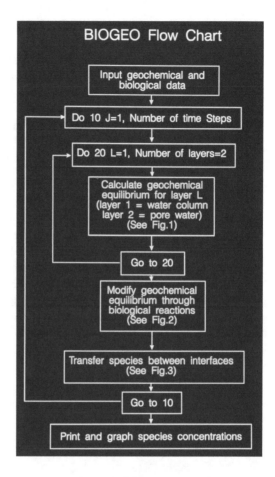

7.2 Charophytes: A Multitasking Tool in Alkaline Water

Charophytes are a taxonomic group of green algae which have an extensive geological history. This is because calcium carbonate is precipitated on the cell walls, making them prominent in the fossil record. Charophytes are considered ancestors of land plants and have attracted substantial scientific interest in the past two decades for unraveling the adaptation to terrestrial plant life (Apolinarska et al., 2011). The fact that these algae have been well studied, has led to a massive bibliography (Van Raam, 2008).

Charophytes are morphologically complex algae, belonging to the Streptophyta (Jeffrey 1967; Stewart & Mattox, 1975). Their large cells have made them model systems for the elucidation of plant cell organization (Foissner & Wastenays, 2014). The large internodal cells have been studied by plant physiologists and physicists since, at least 1974, when Zimmermann & Dainty (1974) defined the membrane transport system in plants. They have very peculiar physiological as well as morphological characteristics which make them particularly well-suited for applications of wastewater treatment and nutrient sinks.

Charophytes enhance the conversion of atmospheric carbon dioxide (CO_2) to solid carbonate minerals as an indirect effect of photosynthesis and deposit the minerals in the form of carbonate crusts (i.e., $CaCO_3$). The charophyte cell wall calcifies to varying degrees, largely depending on the pond water characteristics. Three levels of carbonate formation are distinguished; external, intra-thallus carbonate precipitation, and internal calcification of oospores (Raven et al., 1986; Leitch, 1991; Anadón et al., 2002; Raven & Beardall, 2003; Kawahata et al., 2013). Calcium carbonate encrustations can account for up to 80 % of biomass dry weight (Pukacz et al., 2016).

Carbonates are precipitated on the charophyte cell wall as a result of bicarbonate and nutrient uptake (Smith & Walker, 1980; McConnaughey & Falk, 1991; McConnaughey & Whelan, 1997). In both cases, local H^+ extrusion acidifies the charophyte surface. Remaining hydroxyls (OH^-) are transported in cytoplasmic streaming within the large cells and are released, forming exterior alkaline zones (Beilby & Bisson, 2012). In alkaline zones of the cell walls, cations and carbonate ions are precipitated as carbonates (i.e., $CaCO_3$) on the plant surface (Fig. 7.3a). Carbonate precipitation can occur spontaneously, as indicated by the reaction $Ca^{2+}(aq) + CO_3^{2-}(aq) \leftrightarrow CaCO_3(s)$. However, the precipitation of carbonates is

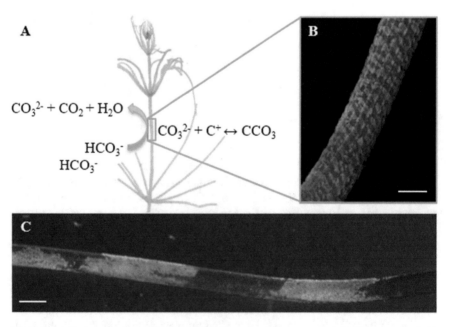

Fig. 7.3 (**a**) Carbon uptake results in precipitation of carbonates on the charophyte surface. C^+ stands for cation, mainly Ca^{2+}, but other cations like K^+, Mg^{2+}, Na^+, and Mn^{2+} are also precipitated. (**b**) Internode of corticated *C. vulgaris*. (**c**) Internode of ecorticated *C. braunii*, scale bar = 500 μm. (Photographs by A. Herbst)

facilitated by *Chara* and linked to the chemical gradients generated during photosynthesis in the vicinity of the cell walls.

On a microscale, in the ecorticated cell walls of charophytes, a regular pattern of acidified and alkalinized zones often becomes visible as a banding pattern on the axes and branchlets (Fig. 7.3b), whereas the banding is less pronounced in corticated species (Fig. 7.3c; Spear et al., 1969; McConnaughey & Falk, 1991; Ray et al., 2003; Kawahata et al., 2013). Most charophytes require waters of relatively high alkalinity. According to Stroede (1937; cited in Dambska, 1964), the minimum calcium concentration for *Chara* appears to be between 15 and 60 mg Ca L^{-1} (Kufel & Kufel, 2002). Generally, species of the genus *Chara* prefer waters higher in Ca content than those of the genus *Nitella* (except *N. mucronata*; Kufel & Kufel, 2002). Charophytes have been reported to have a higher affinity to HCO_3^-, although most algae prefer CO_2.

Charophytes are found in alkaline pH waters, where the dissolved CO_2 is generally absent or low for growth, i.e., photosynthesis. As a carbon source, bicarbonate (HCO_3^-) is abundant and *Chara* has evolved to utilize it to sustain its photosynthesis and primary production. The uptake of carbon by *Chara* is controlled by carbonic anhydrase embedded in the cell membrane. This enzyme catalyzes the transformation of HCO_3^- to CO_2 and facilitates the assimilation of CO_2 required for photosynthesis (Ray et al., 2003). In the process, H^+ concentration increases at the cell-water interface and acidifies water to levels that keep CO_2 in solution and, hence, keep feeding the photosynthesis cycle. Membrane transport mechanisms facilitate the direct uptake of HCO_3^- from water which is followed by intracellular conversion to CO_2. This ability is a unique characteristic of charophytes. With the OH^- which is exported from cell, the pH at the cell-water interface increases, and this favors the precipitation of carbonates (McConnaughey, 1991; Kufel and Kufel, 2002; Ray et al., 2003; Pełechaty et al., 2013; Pukacz et al., 2014; Herbst et al., 2018a; Pertl-Obermeyer et al., 2018; Sand-Jensen et al., 2018).

7.3 Charophytes as Carbon Sinks

Asadian et al. (2018) have discussed the possible use of oceanic algae as indicators of climate change and a partial solution to carbon sequestration. They suggested that algal production in the oceans could transfer carbon from the surface to sediments in the deep ocean, where they would be locked away. Much in the same way, algal production in freshwater ponds and lakes, but especially in mine waste water ponds, might transfer carbon from surface waters to sediments. Charophytes are a very ancient group of algae, as their fossil record reaches back as far as the middle Palaeozoic. Their calcium carbonate encrusted thalli have remained in sediments for thousands of years, suggesting that they are a stable and long-term sink for atmospheric CO_2 (Apolinarska et al., 2011; Pełechaty et al., 2013). Their primary production in mine waste water (and other freshwater ponds and shallow lakes) would not only clean the water but provide a carbon sink. For this, mining companies might be able to both clean their mine waste waters, and potentially obtain carbon credits.

The calcification rate depends on seasonality and limnological factors (e.g., water depth) and how these impact photosynthesis. Yet, ideal conditions for maximum calcification have not been determined since these can be very different among *Chara* species (Pukacz et al., 2016; Herbst et al., 2018a, b). Andersen et al. (2019) highlight the importance of the carbon pump to primary productivity in shallow lakes, and mine waste water ponds can be considered shallow lakes. Nevertheless, average carbonate deposition based on the *Chara* biomass (1725 ± 293 g.m^{-2}) generated in a 6-month period on the littoral zone of a shallow lake (1 m depth) has been estimated as 1255 ± 278 grams of CaCO$_3$ per m^2 (Pukacz et al., 2016). Using these data, with the knowledge that CaCO$_3$ contains 43.97% CO$_2$, and *Chara* biomass generates insoluble CaCO$_3$, it can be inferred that 550 g of CO$_2$ can be sequestered by *Chara* encrustations per m^2. For an area of 1 ha, *Chara* encrustations could retain as much as 5.5 t of CO$_2$ over 6 months. Sand-Jensen et al. (2018) suggested that carbonates remain on cell membranes even after microbial degradation of the organic carbon. He reported that during one summer, the standing biomass of 1000 g.m^{-2} was found to have 59% to over 76% of its dry weight as carbonate, on average 438 g.m^{-2} to 685 g.m^{-2}.

If both calcium and carbonates are present in water, the Saturation Index SI = log$_{10}$(IAP/Ksp) can be estimated. This index relates the ion activity product (IAP)

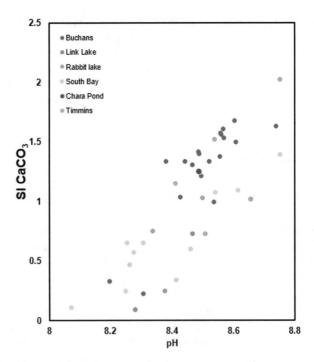

Fig. 7.4 CaCO$_3$ saturation index (SI) for mine waters at different Canadian mines calculated with PHREEQC interactions. (Parkhurst & Appelo, 2013). Indices above one (1) indicate the potential to precipitate

(e.g., Ca^{2+} x CO_3^{2-}) to the solubility product (K_{sp}) of the mineral. As a basic rule, when SI > 0 the precipitation of carbonates is favorable, but if SI < 0 then mineral dissolution occurs. Simply put, the water needs to contain sufficient calcium in the ionic form Ca^{2+}, and not, for example, in the form of gypsum, $CaSOx$. $2H_2O$ or other sulfates. Mine waste waters with charophytes have been examined with respect to the calcite saturation index (SI_{CaCO3}). The results are presented in Fig. 7.4.

All the effluents used in Fig. 7.4 have the chemical conditions necessary to promote calcite precipitation (SI > 0), and, hence, could support cultivation of *Chara* as a carbon sequestration resource. Perhaps *Chara* and other charophytes, if grown in mine waste waters and other shallow lakes and ponds, might just make a significant carbon sink that would allow mines to claim carbon credits. Using ecological engineering designs that account for both an ecosystem's needs and the geochemistry of carbonates might, if scaled-up, make a contribution to the mining industry's need to reduce its carbon footprint.

Not only charophytes but other algae and microbes can precipitate calcium carbonate (Altermann et al., 2006; Diaz & Maberly, 2009; Kamennaya et al., 2012; Zhu & Dittrich 2016). In phytoplankton, coccolithophorids precipitate calcium carbonate on their theca, (Paasche 1968; Moheimani et al., 2012), and coralline algae (Goreau, 1963), cyanobacteria (Dittrich & Sibler, 2010), and filamentous algae, can precipitate calcium carbonate on their thalli. Radionuclides can also be co-precipitated (Kalin et al., 2011; Dienemann et al., 2002; Jha et al., 2010). Many algae form mat-like structures that precipitate calcium carbonate and metals, such as stromatolites (Riding, 2000).

Co-precipitation of metals with calcium carbonate is also a well-documented means of metal sequestration (Hutchinson, 1975; McConnaughey, 1991; Gomes & Asaeda, 2009, 2013). Consequently, there is a large literature on bioremediation using charophytes (Lacerda et al., 1992; Marquardt & Schubert, 2009; Clabeaux et al., 2013; Sooksawat et al., 2013, 2016, 2017). But, what metals and other contaminants can be removed from effluents using charophytes?

The effluent from waste rock piles and the mill at the top of a drainage basin in northern Saskatchewan contained elevated concentrations of radium and uranium. These effluents passed through two lakes separated by a narrow wetland. The concentration of ^{226}Ra increased further as it was released from sediments in the upper of the two lakes. The upper lake in the drainage basin was void of underwater vegetation but the lower lake had a dense cover of the charophyte, *Nitella flexilis*. Regulatory agencies requested the company construct a chemical treatment plant in the narrow muskeg-covered narrows between the two lakes.

The water monitoring data collected by the company revealed significant decreases in ^{226}Ra concentrations as water passed through the muskeg-covered narrows (Fig. 7.5a). Boojum was requested to assess if one of the 'Boojum weeds' might be growing in the wetland. A rather adventurous trip in the winter ($-40°$ C) revealed a large stand of *Nitella flexilis*, which when sampled and exposed to the air, froze instantly to the glove. Samples were placed into plastic bags and submitted for analysis. The radiological analysis revealed high concentrations of ^{226}Ra.

Fig. 7.5 Experimental transplants of *Nitella* at a uranium mine in northern Saskatchewan. (**a**) Photograph shows the lake system downstream from mill building (white dot at horizon). The lower lake is in center, upper lake in the background right corner. The wetland in the foreground and the lower lake were populated by *Nitella* that were removing ^{226}Ra. (**b**) *Nitella* plants in containers ready for transport and transplantation (Boojum Research, 2003). (Photographs by Boojum Research)

A project was launched to test the possibility that these charophytes (*Nitella flexilis*) could be transplanted and grown in the upper lake in the areas where the ^{226}Ra was diffusing through the sediment. The lake was barren of an aquatic vegetation, destroyed during pit construction to reach the ore body (Fig. 7.5a). If the transplantation were successful, it could replace the proposed chemical treatment plant. Biomass in significant quantities was transplanted by helicopter to the upper lake (Fig. 7.5b). The transplants took hold and removed radium and uranium with time (Boojum Research, 2003). The project was completed by 2004, at which time a complete underwater meadow covered the lake sediments.

After several years, the upper lake was extensively populated, and no chemical treatment plant was constructed for the removal of ^{226}Ra. In 1998, process water from the mine was discharged to the upper lake, dramatically altering the water quality. It was feared that the transplanted underwater meadow might not survive. However, as of 2014, no chemical treatment plant had been constructed (J. Jarrell, personal communication, 2014). A final report was issued in 2004 and is available in the report library at Laurentian University (Boojum Research, 2003).

7.4 Algal Blooms: Unchartered Tools for Biological Polishing

Since Boojum Research was established in the early 1980s, its employees have visited many mine sites around the world and observed unusual and interesting interactions between algal assemblages and different ores. It is postulated that many of these algal communities could be scaled-up to improve the quality of the mine effluents associated with these ores. None of the observations below have been followed up, but might, with the right support, further the collection of natural effluent

Fig. 7.6 (**a**) Acid drainage seepage (pH 2–3) found in an adit at a small household coal mine dump in China. A shovel is placed for scale. (**b**) A dense carpet of *Ulothrix* sp. growing in uranium mine effluent in northern Saskatchewan, Canada (pH 7–8). (Photographs by Boojum Research)

cleansing technologies to be used to provide sustainable, long-term solutions for undesirable mine effluents. They are interesting, and might stimulate both mining industry and the scientific communities to develop the ecological tools further.

Example 7.1   In China, Boojum found iron-encrusted algae and fungi growing in a small 'waterfall' area exiting the adit of 'household' coal mine (Fig. 7.6a), where the drainage had a pH of 2–3. This coal is contaminated with antimony and iron, which appeared in the drainage. The algae, fungi and microbes in this small seepage (see shovel for scale) might be the key to antimony removal, although this has not been tested. At another mine in China a system was set up to promote iron and antimony removal with iron precipitation (Sun et al., 2015). The oxidation and precipitation of the reduced iron emerging from the adit can be accomplished by running the effluent over a series of cascading waterfalls, followed by further polishing in algal ponds.

Example 7.2   In northern Saskatchewan, Boojum observed a nearly solid, floating, carpet of *Ulothrix*, a green alga, growing in the narrows between two lakes in the path of uranium mine effluent (Fig. 7.6b). The mine's environmental staff reported that this bloom occurred every year, sometimes thick enough "to walk on." What conditions lead to this phenomenon? Boojum was using another alga, *Nitella*, at this site for removal of radium and uranium, but these other species might be an even better biological polishing tool.

Examples 7.3 and 7.4   Some algal blooms depicted are a 'natural' occurrence throughout the year, or 'bloom' only in spring and fall. Fig. 7.7a shows *Oscillatoria* (a cyanobacterium) growing in a neutralization pond on tailings in acid run-off in northern Ontario. The algae form precipitate complexes which contain, in this case, 17% iron and 7% zinc growing in water concentrations of 0–1 mg.L^{-1} Fe and 10–50 mg.L^{-1} Zn (Kalin & Wheeler, 1992). In another area of the mine, Boojum found

Fig. 7.7 (**a**) A bloom of *Oscillatoria* (brown, floating cyanobacterial mat) in a former neutralization pond located on tailings in northern Ontario. Analyses of the algal-precipitate complexes showed them to contain 17% iron and 7% zinc (Kalin & Wheeler, 1992). (**b**) *Microspora,* a green, periphytic algal bloom growing in an acidic runoff area of a mill pond after iron has precipitated (northern Ontario). (Photographs by Boojum Research)

Microspora growing in very acidic run off from the mill of the same mine. These algae contained 29% iron, but only 0.3% zinc (Fig. 7.7b).

Example 7.5 Boojum investigated two potash mining ponds in Saskatchewan, Canada (Fig. 7.8a). These potash ponds have extremely high pHs, salt and carbonate concentrations. The grayish, darker areas in the pond might be cyanophytes known to assist in desalination. Sediment and water were collected from an evaporation pond. Boojum inoculated hypersaline sediments and found that pink halophytes grew (Fig. 7.8b). The halophytes were coated with a gel-like brownish coating within several weeks. The concentrations of SO_4, Mg Ca, K, Na, and Cl are given in grams and hence represent supersaturated solutions. The algal biomass collected from pond K2 had nearly half of the salt content of the brine in which the algae grew (Fig. 7.8c).

The scientific literature on the biology of halophytes is extensive (Waisel, 2012). Halophytes are also known to be used for the production of biodegradable plastic (Bhati, 2019). Most of the halophytic microbes and fungi which inhabit these soda lakes are cyanobacteria (Lanzén et al., 2013; Krienitz et al., 2016). One of the most common cyanobacterial genera is *Spirulina* (Gimmler & Degenhard, 2001), which, along with *Dunalliela salina,* only occur in alkaline water (Seckbach, 2007; Grant & Sorokin, 2011). *Spirulina* can be cultivated in large quantities for food (Sánchez et al., 2015). Perhaps some of these other algae will assist with desalinization?

Example 7.6 At a mine in northern Queensland, Australia, at that time one of the largest zinc mines in the world, magnesium and sulfate concentrations exceeded regulatory limits during the monsoon rains. The drainage from the waste rock piles was collected in an evaporation pond. Along the beach, interesting structures were noted, appearing somewhat like stromatolites or tufa (Fig. 7.9a; Freytet & Verrecchia, 1998). All sticks of wood in the water were covered with a similar brown crust. The bottom of the drainage creek was covered by an extensive growth of mostly *Ulothrix* sp*., Zygnema* sp., and *Navicula* sp. *(*Fig. 7.9b*)*. The white, float-

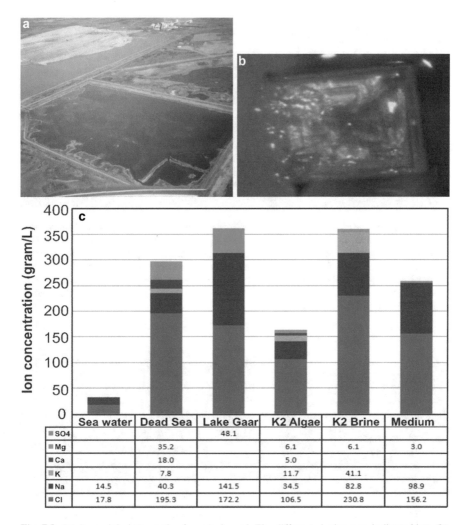

Fig. 7.8 (**a**) An aerial photograph of a potash pool. The different shades may indicate biota (**b**) Photograph of pink halophyte cultures. The colonies were covered with gel-like substances within weeks of culture. (**c**) Comparison of salinity of potash evaporation ponds and natural saline environments. The concentrations of SO_4, Mg Ca, K, Na, and Cl are given in grams and hence represent supersaturated solutions. The algal biomass collected from pond K2 had nearly half of the salt content of the brine in which the algae grew. (Photographs by Boojum Research)

ing material in the creek was analysed and contained an average of 11.6% Ca, 12.0% S and 2.5% Mg (Fig. 7.9b). Extrapolating this removal ability to a one-hectare pond (if covered with the crusts) would extract 1.0 tonnes of Ca, 1.1 tonnes of S and 0.22 tonnes of Mg. It follows that if the regulated elements, S and Mg, are soluble, they could be recycled. But, if insoluble, they might be removed, if the pond were skimmed of the crusts.

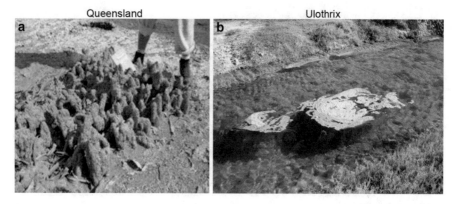

Fig. 7.9 (a) Alkaline biostructures along the entire beach of an evaporation pond at a zinc mine in northern Queensland, Australia. Structures are formed during monsoon season. The structures initiated on wooden sticks (not shown) which get covered with a living, brown coating similar in color to the crusts shown. These crusts resemble tufa or stromatolite structures. (b) Photograph of the creek carrying the drainage from the waste rock pile along with the floating material. Note the green periphyton along the creek bottom, and the white precipitate floating on the surface. (Photographs by Boojum Research)

Fig. 7.10 (a) General view of a well-vegetated valley seep pond. (b) Algal growth in a natural waterfall on the side of one of the valley seep ponds, a possible further means to remove selenium. (Photographs by Boojum Research)

Example 7.7 A feasibility study was conducted assessing the applicability of ecological engineering to lower selenium concentrations in drainage collection ponds from mountain-top coal mining in West Virginia, USA. The conventional approach to water treatment is ponding of the drainage, where biological methylation in the sediment and by the vegetation leads to a reduction of selenium through volatilization to the air (De Souza et al., 2002). When the ponds are filled with sediments, they are removed to a landfill site, and the ponds refill gradually. Because the ponds are alkaline, and charophytes are abundant in many of them, Boojum was retained to assess the potential of using these plants to sequester and remove selenium.

All ponds visited were invaded with emergent and submergent aquatic plants, including charophytes (Fig. 7.10a). As expected, the selenium concentrations leaving the pond were somewhat lower than those entering the pond, but still not low enough for the regulators. It was suspected that the sediments accumulated selenium (Schaivon & Pilon-Smits, 2017). If this was the case, selenium may not only be accumulated, but recycled from the sediments into the plants over the following season.

Sediment cores confirmed a gradient of selenium enrichment; the upper layer (0–2 cm) contained 17 µg.g^{-1}, 7 µg.g^{-1} at a depth of 2–4 cm, and 1.37 µg.g^{-1} at 4–6 cm. The sediment at 6 cm depth had concentrations of only 0.6 µg.g^{-1}. The selenium concentration in *Chara* biomass ranged from 2.2 to 8.0 µg.g^{-1}. Water was sampled where the inflow and outflow of the ponds could be accessed. When charophytes were sparingly present, the outflow water was reduced to 7 µg.L^{-1}. *Chara* biomass removed between 0.07 g.m^{-2} in sparsely colonized ponds, but up to 0.9 g. m^{-2} in ponds that were more densely populated.

One pond was totally dominated by charophytes and the concentration of selenium at the outflow was below the detection limit. Unfortunately, no boat was available to collect water at the pond inflow. The pre-feasibility study of the valley drainage ponds indicated that, if *Chara* were introduced and flourished in the pond after sediment removal, selenium concentrations in the discharge of the ponds would be significantly lower. A further means of selenium removal might be discovered if the attached algae growing in the 'waterfall' above the pond were investigated (Fig. 7.10b).

7.5 Magnesium Alloys to Support Biological Polishing

Many abandoned or orphaned mines exist where iron oxidation has progressed beyond a pH value of 2 forming a ferric sulfate leach solution, which has been shown to dissolve the mineral matter and trace element content of several coals (Hamersma et al., 1977; Meyers, 1977). Boats in these waters have shiny aluminum bottoms.

Several neutralization options can be used as an alternative to lime additions. For example, the corrosion of magnesium metal (Mg) in water consumes hydrogen ions, raising the pH. The following reactions are probable:

$$Mg^{2+} + H^{2+} \rightarrow Mg(OH)_2 \, (\text{solid}) MgSO_4 \, (\text{water soluble}) + H_2 \, (\text{gas}) \quad (7.1)$$

which might form hydro-magnesite, nesquehonite and/or lansfordite. The carbonates ultimately sink to the sediment, while the magnesium sulfate stays dissolved, but is relatively unreactive. Corrosion of the metallic magnesium continues as long as the pH is low (Song & Atrens, 1999), even though the surfaces become coated with carbonates/hydroxides (Fig. 7.11a). As this reaction rate is slow, it is not detrimental to the existing aquatic biota.

Fig. 7.11 (a) A close-up photograph of corrosion channels with hydrogen bubbles, due to continued corrosion. (b) Magnesium scrap metal on a string covered with iron and magnesium after being suspended from a barge in an acid lake. (Photographs by Boojum Research)

Larger scale experiments were carried out in an acidic lake in central Ontario. Magnesium metal scraps were tied on ropes to barges floating in the lake. When the barges shifted with the wind, the movement of the scrap introduced new lake water to the magnesium, causing hydrogen gas to form small craters within the magnesium carbonate/hydroxide, allowing bubbles to escape (Fig. 7.11b). From 1999 through 2002, a total of 4.6 tonnes of magnesium scrap were added to the lake barges, representing a total surface area of 1339 m^2. The goal was to eventually raise the lake pH to 4.5 with more barges. At this pH, enough CO_2 would remain in solution as bicarbonate to make algal growth more feasible. The amount of magnesium added with the rafts may or may not have been sufficient to raise the lake pH significantly, but it appeared to slow acidification of the lake. In fact, during the winter of 2003, water sampled from below the ice near one of the barges had a pH of 8. This pH was localized around the barge, and would allow for carbonate precipitation, as indicated in Fig. 7.11. As such, a side effect of this neutralization strategy with MgO is the potential to absorb atmospheric carbon as magnesium carbonates, a process that is highly desirable as a means to decrease the industrial carbon footprint (McQueen et al., 2020).

Magnesium is a very common element in nature. It is among the ten most abundant elements in the accessible geosphere. In the earth's crust it commonly occurs as a component of the mineral dolomite $CaMg(CO_3)_2$, kieserite $MgSO_4 \cdot 2\ H_2O$ or olivine $(Mg,Fe)_2SiO_4$. Magnesium is useful in environmental restoration as it is readily attacked by oxygen and water producing non-toxic secondary products. With oxygen, magnesium forms white MgO, a very brittle material that is easily transformed into a powder. With water, magnesium forms $Mg(OH)_2$, a gelatinous hydroxide that, in contact with air, transforms into a variety of mixed oxide/hydroxide/carbonate compounds. In no case are these corrosion products of magnesium metal stable enough to prevent further corrosion. So, it forms no crust or oxide layer on metal surfaces as long as the medium is at least slightly acidic. It is well known that if acid is neutralized with limestone the secondary products are more stable and hence lead to crust formation and ultimately passivation of the reactant and neutralization is no longer effective.

The dissolution of metallic magnesium in water formally results in $Mg(OH)_2$ and H_2

$$Mg(s) + H_2O(aq) \rightarrow Mg(OH)_2 + H_2(g) \qquad (7.2)$$

In acidic waters, transiently $Mg(OH)_2$ is instable (Esmaily et al., 2017) and dissociates

$$Mg(OH)_2 \rightarrow Mg^{2+} + 2\,OH^- \qquad (7.3)$$

where the hydroxide is available for AMD neutralisation. The magnesium cation, however, is invariably divalent and, thus, does not form precipitates that may form inhibiting surface covers.

A major attraction of this AMD neutralisation approach by scrap magnesium is the slow but steady production of hydrogen ions through corrosion of the metal surface *in-situ* under normal atmospheric conditions. Seen chemically, addition of scrap magnesium into the acidic lake water acts like a hydroxide titrant reducing H^+ by neutralisation. The speciality with scrap magnesium is that the titrant is generated comparatively slowly due to heterogeneous reaction at the surface of the solid Mg material. The reaction comes to an end when either the scrap magnesium is consumed or the acidity is sufficiently reduced. The need for human surveillance of the process is minimised.

More or less pure Mg scrap is not available commercially. Available materials are almost always alloyed. The material used in the scale-up experiments was an Mg alloy ASTM AZ91D, holding 9% Al and 1% Zn. The scrap Mg was delivered in 1 m^3 bags at a weight of 300 kg per bag. The alloy is more corrosion-resistant compared to 99.9% Mg but proved to show a satisfactory corrosion rate in all experiments. For initial experiments, though, Mg alloy ASTM AZ31 was provided where a reduced aluminum content of 3% was used. To estimate the impact of alloying on the reactivity of Mg, a few pure Mg blocks had been purchased with a total weight of approx. 10 kg.

7.5.1 Magnesium Alloy Flow-Through Experiment on Lake Shore

The reported large-scale experiments with magnesium scrap metal (Mg or Mg alloy) were carried out in a mining lake in northern Ontario (Canada) between 1998 and 2002. The lake water pH at this time was about pH 3. These experiments aimed to arrest the seasonal decreases in pH due to spring turnover. If stabilization at pH 4.5 was successful, enhanced phytoplankton productivity would boost the biological polishing capacity. The collected geochemical and hydrological information were used for a rough assessment of the magnitude of chemical effects on the lake.

Preliminary laboratory experiments with the acidic lake water (430 mg.L^{-1} of Ca, 4760 mg.L^{-1} of Fe, 210 mg.L^{-1} of Mg, 3950 mg.L^{-1} of S) showed that the dissolution of 0.13 mol MgO in 1 L of lake water could help raise the pH from 3 to pH 9. Typical experimental results are given in Fig. 7.12a together with geochemical modelling results based upon PHREEQC. The neutralisation reaction terminated within 50 h resulting in an increase from pH 3 to pH 9. Fig. 7.12b gives the change in saturation indices of selected Fe^{3+}, Ca^{2+} and Mg^{2+} minerals as a function of pH. An SI below zero indicates undersaturation and, (hence) dissolution of the respective minerals, while an SI above zero suggests precipitation. Fig. 7.12b illustrates that trivalent Fe^{3+} hydroxides readily precipitate at pH 3 and above, while divalent cations like Ca^{2+} and Mg^{2+} precipitate only at pHs above 8. Thus, the formation of inhibiting layers and crusts on the scrap magnesium surfaces are unlikely in even slightly acidic mine waters.

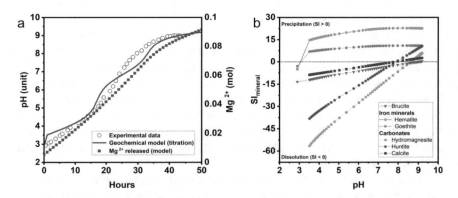

Fig. 7.12 (a) Behaviour of experimental lake water upon addition of scrap magnesium. (a) The experimentally observed change in pH occurred within 50 h. This reaction was modelled by PHREEQC (Parkhurst & Appelo, 2013) with results given as the red curve. The calculated overall increase in Mg^{2+} is also given. Experiment and modelling were performed with 10 mL water. (b) shows results of a model simulation of solubilities of selected Fe^{3+}, Ca^{2+} and Mg^{2+} minerals hematite ($Fe(OH)_3$), goethite ($FeO(OH)$, brucite (MgO), hydromagnesite ($Mg_5(CO_3)_4(OH)_2 \cdot 4H_2O$), huntite ($Mg_3Ca(CO_3)_4$) and calcite ($CaCO_3$). Modelling results are based on the PHREEQC standard database

Fig. 7.13 Field pilot test of Mg alloy surfaces in AMD for defining amount of Mg needed to be suspended to buffer decreasing seasonal pH decline. (**a**) A cascade of three 1 m³ polyethylene containers. (**b**) Second experimental setup. (Photographs by Boojum Research)

The modelling results in Fig. 7.12a,b are given for illustration. The authors are aware of the fact that numerical modelling needs to be seen with caution. Figures 7.12a, b are however, suitable to illustrate the basic chemical principle of AMD neutralisation with scrap magnesium.

The dimensions of these experiments can be appreciated from Fig. 7.13. Batch and flow-through experiments were performed at the shores of the experimental lake during June 1998. These experiments were designed to examine whether H⁺ increases and acidity reductions observed in the laboratory experiments could be repeated in field conditions, and whether the rates of these reactions were comparable to rates determined in the lab.

In these experiments, containers held various amounts of AZ31 scrap magnesium (Fig. 7.13). A 1 m³ plastic tank was filled with experimental lake water. A sack, sewn from fibreglass window screen material, containing 27 g of Mg cuttings was suspended in the tank. This approach was taken in order to provide a minimal mass of Mg in a form with high surface area. The solution was periodically stirred by hand using a 2 in. × 2 in. wooden dowel. At night, the solution was stagnant. The pH of the solution gradually rose from pH 3.2 to pH 3.52 over 3.86 days. When the pH reached about 3.5, the solution became turbid and orange. Overall, the process was very slow.

The Mg plates used in the circulated and stagnant batch tanks were transferred to a single tank, where fresh experimental lake water (pH 3.2) was metered into the tank, and overflow captured in a third tank. This experiment is shown in Fig. 7.13b. The flow-through tank contained the stagnant batch experiment water (pH 5.2) whose pH had been adjusted down to pH 4.1 with fresh feed water. A small flow of water (0.004 to 0.026 L.s⁻¹) was metered into the tank, but flow control was difficult due to the crude adjustment capacity of the valve. The pH of the reactor tank varied over the 2.2-day experiment, ranging from pH 3.9 to 4.7, but did not reach equilibrium, due to variable inflow. In total, approximately 2.2 m³ of experimental lake water passed through the reactor.

While the pH of a bulk solution may be low, the conditions close to the reacting surface may be quite different. Close to the surface where the neutralizing alkalinity

Fig. 7.14 Field
experiments (1 m³) with
exposure of 7.4 kg of scrap
magnesium to circulating
(red) and static (blue)
experimental lake water

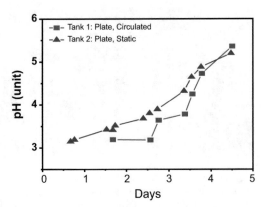

is generated comparatively high pH values may prevail. Thus, geleous $Mg(OH)_2$ may hamper the alkalinity-forming reaction. In order to get a hint of the order of magnitude of such effects, the time-dependent change in pH was studied in circulating and stagnant experimental lake waters. Several kilograms of scrap magnesium were contacted with 1 m³ experimental lake water. A typical result is shown in Fig. 7.14. In this case, the static sample reacted faster compared to the circulating sample. However, taking the poorly controlled conditions (e.g., with respect to the effective surface area of scrap Mg) into account, no significant difference between circulated and static samples could be observed.

7.5.2   Estimating Neutralization Kinetics

Nevertheless, corrosion products attached to the surface of a magnesium particle reduce access of oxidizing agents, e.g., oxygen and/or water. Over-all kinetics of Mg corrosion is a function of the accessible surface, the amount of corrosion agent available for reaction per time unit, the temperature at the corrosion site and the activation energy of the corrosion reaction. In a general way the process may be described by a kinetic equation

$$\frac{\partial\left[Mg(II)\right]}{\partial t} \approx F\,k\left[Kp\right]^{v} \tag{7.4}$$

In Eq. 7.4, the change in the Mg^{2+} concentration with time is expressed by the accessible Mg metal surface per unit volume, F, a general expression of the corroding agent(s), KP, the unknown reaction order v, and the kinetic constant k.

In the absence of detailed information about the reaction order and the detailed chemical nature of the corroding agents, KP, it may be assumed that these parameters will not change drastically with the given conditions. In the present situation, the major corroding agent is water. The magnesium metal is completely immersed in the water body of the lake. Hence, major changes in its concentration are very unlikely. The reaction order depends on the reaction mechanism and may change with temperature as well as with the concentrations of all species involved in the reaction. Because the temperature of the water is not prone to drastic variations and the components involved in the chemical reaction are quickly dispersed and diluted, it may be reasonably assumed that the term $[KP]^\nu$ cannot be influenced to increase reaction rates. Thus, the accessible area F has a considerable influence on the over-all kinetics of the corrosion process of magnesium in water solutions. Since the coatings and crusts formed on the Mg surfaces in lake waters are brittle and often also finely dispersed, movement of the Mg in the water, either by wind and waves or forced by pumping, will have a strong influence on the over-all reaction rate of the process. This conclusion is clearly confirmed by the observations. It might be possible to further improve the kinetics of the process by avoiding the contact of Mg rods in the water with carbonates from air, e.g., by forced circulation using pumps. Here further experimentation may result in a better understanding of the possible benefits.

The kinetic constant may be expressed by an Arrhenius-like relationship (Eq. 7.5)

$$k = Ae^{-\frac{\Delta E^H}{RT}} \tag{7.5}$$

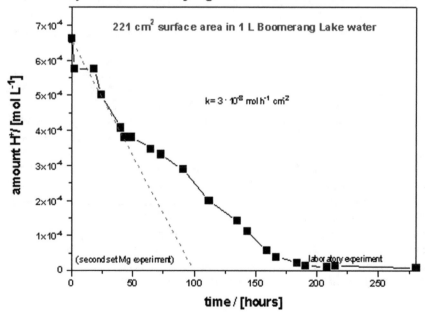

Fig. 7.15 Example of the estimation of an overall kinetic factor k with a defined scrap magnesium surface in 1L experimental lake water

where A is a kinetic pre-factor, ΔE^H is the Gibbs free energy of activation, R is the gas constant and T the absolute temperature in Kelvin. Hence, the kinetic factor k for a given chemical reaction depends essentially on temperature.

A series of experiments were performed under laboratory conditions to get a rough order of magnitude for the acid-neutralisation capacities of scrap magnesium in experimental lake water. A typical example is shown in Fig. 7.15. The onset of hydrogen generation and acid reduction was rapid initially and decreased after several days. The acid reduction rate was sufficiently fast to support the design of large-scale experiments at the site.

Using such information for estimating the likely change induced by the addition of scrap magnesium into a surface water body, the following points need to be considered:

- The estimated overall kinetic factor is based on the initial scrap magnesium surface area. This parameter certainly will change in the course of an experiment. The magnitude of changes will be unknown. Therefore, only the initial slope has been used. 'Initial' was often used in a pragmatic way and ad-hoc manner. The small number of data points often did not leave another approach.
- Influence of temperature seemed to be considerable – as expected by Eq. 7.5. Hence, summer seems to be the preferential time for addition of scrap magnesium into lakes.
- Apparently, movement of scrap magnesium in water seems to be beneficial. It is likely that hydrogen bubbles (cf. Eq. 7.2) may attach to the metal surface. At higher pH values gelatinous $Mg(OH)_2$ may form close to the reaction zone at the metal surface. Both effects will reduce the accessible scrap magnesium surface area. Movement in the water is expected to reduce these adverse effects.

Since temperature at the bottom of shallow lakes is usually quite constant, k varies within narrow limits. On the other hand, Eq. 7.5 may also suggest that keeping the scrap magnesium close to the surface may be beneficial in order to profit from faster reaction at warmer temperatures. As a consequence of these discussions, different configurations for controlled Mg exposure to the lake waters were tested at the Boojum lab. As a result, two configurations have been studied in closer detail: the raft configuration and the box configuration. The critical evaluation of the available data made clear that the box configuration was preferable. The parameters determined by Boojum for both systems will be summarized in terms of the above stated reaction kinetics. Ongoing experiments, in combination with an additional forced flow-through experiment, which were carried out in the laboratory by Boojum, determined the parameters of the above given kinetic equations for the boundary conditions of the lake water. The goal was an over-all estimate of the amount of Mg metal and number of boxes necessary to achieve controlled increase in pH in the experimental lake with optimal economic resources.

7.5.3 Magnesium Alloys in Rafts and Barges in an Experimental Lake

Supported by results from the 1 m³ tests (cf. Fig. 7.13), a total of 4.6 tonnes of magnesium scrap were added to the acidic lake between 1999 and 2002. Four separate additions of scrap magnesium (April 6, 1999: 970 kg and 2931 kg; June 28, 2001: 402 kg; Sept 20, 2002: 324 kg) totaled 4.6 metric tonnes. These 4.6 tonnes were added to a volume of about 1,000,000 m³ lake water. Furthermore, an average annual replacement of lake water on the order of 300,000 m³ was estimated. This implies that the lake water is completely replaced every three years. These represent only rough estimates. A more detailed hydrological model of the lake would be necessary to account for partial mixture, the dilution effects of the generated Mg^{2+} ions and associated increase in pH. Such a model requires more detailed hydrological investigation and is outside the scope of this study which concentrated on the feasibility of acid reduction by addition of scrap magnesium.

The scrap magnesium was suspended in rafts and barges (Fig. 7.16) that were positioned in different locations within the lake. These materials were estimated to have an exposed area for reaction equivalent to 1339 m² (Table 7.1).

In this period, the dissolved Mg concentrations in the lake water increased from 10–12 mg.L⁻¹ to 16–18mg.L⁻¹ (with seasonal variations). Corrosion rates were much faster in open water due to greater water circulation and supply of H⁺ to MgO surface. Water sampled near the immersed Mg-scrap metal in the winter, when the

Fig. 7.16 Tests using the corrosion of magnesium metal to raise lake pH. (**a**) wooden raft containing compact magnesium alloy pieces which could not be fastened with a string. (**b**) barge set-up with alloy pieces suspended on ropes. (Photographs by Boojum Research)

Table 7.1 Magnesium placement in the experimental lake

	Date Mg added	Location in Lake	Weight in kg	Surface Area in m²	Ratio kg.m⁻²
Rafts	06-Apr-99	in quiet bay	3901	1047	3.73
Barge 1	28-Jun-01	in open lake	264	133	1.98
	20-Sep-02		138	47	2.91
Barge 2	20-Sep-02	shallow water (close to outflow)	325	112	2.91

barge was frozen in place, had a pH of 8. Numerous precipitates were observed on corroded MgO surfaces (Fig. 7.11a, b). The lake water was charged with high concentrations of iron, calcium, magnesium, and sulfur, allowing neutralization reactions and precipitation to proceed on the MgO surface. These precipitates, most probably iron hydroxide (oxides and potentially carbonates) formed coatings and crusts on the Mg surfaces. However, when the barges were visited, the bubbles emerged and shaking the barge the coating fell off to the sediment. The overall reaction rates are dependant on the movement of the barge and the exposure of scrap magnesium surface.

The field tests caused an increase in Mg concentrations and pH near the immersed metal (i.e., closer to reactive surfaces) that strongly suggested that full-scale implementation of the Mg scrap technology can be an alternative to current neutralization techniques. Moreover, if carbonate precipitation can be simultaneously promoted and maintained, another positive long-term effect of the scrap magnesium application to acid mine drainage would be atmospheric CO_2 sequestration. However, the amount of scrap magnesium added to the lake was not sufficient to cause a drastic shift in the lake pH.

Theoretically, the 4.6 tonnes of magnesium added to the lake will consume 258,711 moles of H^+. Assuming a k value an order of magnitude lower under field conditions, averaging winter and summer temperatures and allowing for insufficient agitation, a pH increase in the experimental lake from 3.0 to 3.1 would require that 258,690 moles of H^+ would have to be consumed. However, these changes are hardly measurable. To raise the pH from 3 to 4 in an acidic lake with a 10^9 L volume, would require the elimination of 900,000 moles of H^+. Since each mole of Mg^{2+} ion consumes two moles of H^+, this would require an increase in Mg metal (alloy) to between 11 and 18 tonnes.

In April 1999, eight floating rafts with 3.9 tonnes of scrap magnesium threaded on rope and suspended from the rafts were anchored in a small bay (Fig. 7.16). The scrap magnesium strands were enclosed with a plastic curtain to facilitate pH measurements. As it happened, the curtains inhibited the agitation of the magnesium by wave action, a key factor in reaction rates; they were therefore not used on the barges that were subsequently utilized to suspend open magnesium baskets. Moreover, the barges were placed in open water to expose them to more agitation by waves.

Measurements of pH in the vicinity of the Mg rafts did not show promise. The reaction of the magnesium was slow – although it was still occurring four years later. A faster reaction is expected from the Mg suspended in baskets from two barges in open water. In June 2001, a barge supporting 264 kg of scrap Mg pieces was anchored near the experimental lake outflow (Fig. 7.11b). In September 2002, an additional 138 kg of scrap magnesium pieces were added to the barge and a second barge, supporting 325 kg of scrap magnesium was put into the lake. In the winter of 2002, a value of pH 8 was recorded in the magnesium baskets.

These exploratory experiments suggest a few conclusions:

- Two observations are central for the preliminary assessment. The first is the observed increase in Mg ions over the experimentation period. The second observation is the lake water pH which essentially remained rather stable at around pH 3. From the mass balance an observable increase in pH of the lake water was not expected.
- The laboratory measurements which allow a more stable control of the measurement environment, indicate a tendency towards higher acidities (pH < 3) at the end of the experimentation period (exhaustion of scrap magnesium and consequent cease of H^+ neutralisation capacity).
- The measurements of Mg content in the lake water, performed during the period 1987–2002, indicate a significant increase in the Mg content from 10–12 mg.L^{-1} to 16 – 18 mg.L^{-1}(with seasonal variations). Since the total water content in the lake is about 10^9 L, the calculated dissolved amount of Mg is in the order 4 tonnes to 8 tonnes. This estimate is reasonable considering the list of approximations.
- To increase the pH from pH 3 to pH 4 in a lake with 10^9 L^{-1} of water – ignoring flow-in and flow-out – requires the elimination of 9×10^5 mol of H^+. Since each formula unit of Mg^{2+} ion consumes two formula units of H^+, $4.5 \cdot 10^5$ mol of Mg are required, amounting to about 11 tonnes of Mg. Due to flow-in and flow-out, the effect of the actually added 4.6 tonnes is further reduced with the consequence that an effect of the dissolved Mg on the lake pH is hardly visible (but documented via the Mg^{2+} increase).
- Taking the effects together, considering that application of Mg in a full-scale experiment will occur simultaneously at different positions of the lake in a much shorter period, and considering that the most effective method of interaction between lake water and Mg scrap (as evaluated from the exploratory experiments) will be used exclusively, it seems not unreasonable to estimate the total amount of scrap Mg to 16 tonnes. A certain amount of surplus is included which accounts for dilution effects, reduction in surface with progressing reaction of Mg, acidic water, and formation of corrosion covers on the surface of the scrap Mg.

References

Aharchaou, I., Bahloul, F., & Fortin, C. (2020). Competition among trivalent elements (Al, Eu, Fe, Gd, Nd, Tm, and Y) for uptake in algae and applicability of the biotic ligand model. *Archives of Environmental Contamination and Toxicology*. https://doi.org/10.1007/s00244-020-00786-z

Altermann, W., Kazmierczak, J., Oren, A., & Wright, D. T. (2006). Cyanobacterial calcification and its rock-building potential during 3.5 billion years of Earth history. *Geobiology, 4*(3), 147–166. https://doi.org/10.1111/j.1472-4669.2006.000.x

Anadón, P., Utrilla, R., & Vázquez, A. (2002). Mineralogy and Sr – Mg geochemistry of charophyte carbonates: A new tool for paleolimnological research. *Earth and Planetary Science Letters, 197*(3–4), 205–214. https://doi.org/10.1016/S0012-821X(02)00481-8

Andersen, M. R., Kragh, T., Martinsen, K. T., Kristensen, E., & Sand-Jensen, K. (2019). The carbon pump supports high primary production in a shallow lake. *Aquatic Sciences, 81*(2), 24. https://doi.org/10.1007/s00027-019-0622-7

Apolinarska, K., Pełechaty, M., & Pukacz, A. (2011). $CaCO_3$ sedimentation by modern charophytes (*Characeae*): Can calcified remains and carbonate $\delta^{13}C$ and $\delta^{18}O$ record the ecological state of lakes? A review. *Studia Limnologica et Telmatologica, 5*, 55–66.

Asadian, M., Fakheri, B. A., Mahdinezhad, N., Gharanjik, S., Beardal, J., & Talebi, A. F. (2018). Algal communities: An answer to global climate change. *Clean – Soil, Air, Water, 46*(10), 1–14. https://doi.org/10.1002/clen.201800032

Baker, B. J., Lutz, M. A., Dawson, S. C., Bond, P. L., & Banfield, J. F. (2004). Metabolically active eukaryotic communities in extremely acidic mine drainage. *Applied and Environmental Microbiology, 70*(10), 6264–6271. https://doi.org/10.1128/AEM.70.10.6264

Barker, W. W., & Banfield, J. F. (1998). Zones of chemical and physical interaction at interfaces between microbial communities and minerals: A model. *Geomicrobiology Journal, 15*(3), 223–244. Retrieved from: https://doi.org/10.1080/01490459809378078

Beilby, M. J., & Bisson, M. A. (2012). pH banding in charophyte algae. In A. Volkov (Ed.), *Plant electrophysiology, methods and cell electrophysiology* (pp. 247–271). Springer. https://doi.org/10.1007/978-3-642-29119-7_11

Bhati, R. (2019). Biodegradable plastics production by cyanobacteria. In M. Khoobchandani & A. Saxena (Eds.), *Biotechnology products in everyday life. EcoProduction 2018 (Environmental issues in logistics and manufacturing)*. Springer. https://doi.org/10.1007/978-3-319-92399-4_9

Boojum Research (2003). Ecosystem restoration in the Rabbit Lake drainage basin: Retaining 226Ra and uranium within the waste management Area. Report prepared for Cameco Corp. Retrieved from: https://zone.biblio.laurentian.ca/handle/10219/2916

Breuker, A., Ritter, S. F., & Schippers, A. (2020). Biosorption of rare earth elements by different microorganisms in acidic solutions. *Metals, 10*(7), 1–14. https://doi.org/10.3390/met10070954

Buffle, J., & van Leeuwen, H. P. (1992). *Environmental particles* (Vol. 1, p. 576). Lewis Publishers.

Casado-Martinex, C. (2013). Oekotoxzentrum: Ecotoxicity of rare earth elements: Info sheet EAWAG. https://www.ecotoxcentre.ch/media/40675/2013_selteneerden_en.pdf

Clabeaux, B. L., Navarro, D. A., Aga, D. S., & Bisson, M. A. (2013). Combined effects of cadmium and zinc on growth, tolerance, and metal accumulation in *Chara australis* and enhanced phytoextraction using EDTA. *Ecotoxicology and Environmental Safety, 98*, 236–243. https://doi.org/10.1016/j.ecoenv.2013.08.014

Crist, R. H., Oberholser, K., Schwartz, D., Marzoff, J., Ryder, D., & Crist, D. R. (1988). Interactions of metals and protons with algae. *Environmental Science & Technology, 22*(7), 755–760. https://doi.org/10.1021/es00172a002

Dąmbska, I. (1964). *Charophyta-Ramiencie*. Państwowe Wydawnictwo Naukowe.

De Souza, M. P., Pickering, I. J., Walla, M., & Terry, N. (2002). Selenium assimilation and volatilization from selenocyanate-treated Indian mustard and muskgrass. *Plant Physiology, 128*(2), 625–633. https://doi.org/10.1104/pp.010686

Diaz, M. M., & Maberly, S. C. (2009). Carbon-concentrating mechanisms in acidophilic algae. *Phycologia, 42*(2), 78–85. https://doi.org/10.1108/17506200710779521

Dienemann, C., Dudel, G. E., Dienemann, H., & Stolz, L. (2002). Retention of radionuclides and arsenic by algae downstream of U mining tailings. In B. J. Merkel, B. Planer-Friedrich, & C. Woldersdorfer (Eds.), *Uranium in the aquatic environment* (pp. 605–614). Springer.

Dittrich, M., & Sibler, S. (2010). Calcium carbonate precipitation by cyanobacterial polysaccharides. Special Publication #336. In H. M. Pedley & M. Rogerson (Eds.), *Tufas and Speleotherms: Unravelling the Microbial and Physical Controls* (pp. 51–63). https://doi.org/10.1144/SP336.4

Esmaily, M., Svensson, J. E., Fajardo, S., Birbilis, N., Frankel, G. S., Virtanen, S., Arrabal, R., Thomas, S., & Johansson, L. G. (2017). Fundamentals and advances in magnesium alloy corrosion. *Progress in Materials Science, 89*, 92–193. https://doi.org/10.1016/j.pmatsci.2017.04.011

Foissner, I., & Wastenays, G. (2014). Chapter seven – Characean internodal cells as a model system for the study of cell organization. *International Review of Cell and Molecular Biology, 311*, 307–364. https://doi.org/10.1016/B978-0-12-800179-0.00006-4

Freytet, P., & Verrecchia, E. P. (1998). Freshwater organisms that build stromatolites: A synopsis of biocrystallization by prokaryotic and eukaryotic algae. *Sedimentology, 45*(3), 535–563. https://doi.org/10.1046/j.1365-3091.1998.00155.x

Gale, N. L., & Wixson, B. G. (1979). Removal of heavy metals from industrial effluents by algae. *Developments in Industrial Microbiology, 20*, 259–273.

Gimmler, H., & Degenhard, B. (2001). Alkaliphilic and alkali-tolerant algae. In *Algal adaptation to environmental stresses* (pp. 291–321). Springer.

Gomes, P. I. A., & Asaeda, T. (2009). Phycoremediation of Chromium (VI) by *Nitella* and impact of calcium encrustation. *Journal of Hazardous Materials, 166*, 1332–1338. https://doi.org/10.1016/j.jhazmat.2008.12.055

Gomes, P. I. A., & Asaeda, T. (2013). Phytoremediation of heavy metals by calcifying macroalgae (*Nitella pseudoflabellata*): Implications of redox insensitive end products. *Chemosphere, 92*, 1328–1334. https://doi.org/10.1016/j.chemosphere.2013.05.043

Goreau, T. F. (1963). Calcium carbonate depositions by coralline algae and corals in relation to their role as reef builders. *Annals of the New York Academy of Sciences, 109*(1), 127–167. https://doi.org/10.1111/j.1749-6632.1963.tb13465.x

Grant, W. D., & Sorokin, D. Y. (2011). Distribution and diversity of soda lake alkaliphiles. In *Extremophiles handbook* (pp. 27–54).

Hamersma, J. W., Kraft, M. L., & Meyers, R. A. (1977). Applicability of the Meyers process for desulfurization of U.S. Coal (A Survey of 35 Coals). In *Coal desulfurization* (Vol. 64, pp. 11–143). American Chemical Society. https://doi.org/10.1021/bk-1977-0064.ch011

Herbst, A., Henningsen, L., Schubert, H., & Blindow, I. (2018a). Encrustations and element composition of charophytes from fresh or brackish water sites–habitat-or species-specific differences? *Aquatic Botany, 148*, 29–31.

Herbst, A., von Tümpling, W., & Schubert, H. (2018b). The seasonal effects on the encrustation of charophytes in two hard-water lakes. *Journal of Phycology, 54*(5), 630–637.

Hutchinson, G. E. (1975). *A treatise on limnology vol 3, Limnological botany* (p. 660). Wiley.

Jeffrey, C. (1967). The origin and differentiation of the archegoniate land plants: A second contribution. *Kew Bulletin, 21*(2), 335–349.

Jha, V. N., Tripathi, R. M., Sethy, N. K., Sahoo, S. K., Shukla, A. K., & Puranik, V. D. (2010). Bioaccumulation of ^{226}Ra by plants growing in freshwater ecosystem around the uranium industry at Jaduguda, India. *Journal of Environmental Radioactivity, 101*(9), 717–722. https://doi.org/10.1016/j.jenvrad.2010.04.014

Juwarkar, A. A., Singh, S. K., & Mudhoo, A. (2010). A comprehensive overview of elements in bioremediation. *Reviews in Environmental Science and Biotechnology, 9*(3), 215–288. https://doi.org/10.1007/s11157-010-9215-6

Kalin, M., & Wheeler, W. N. (1992). A study of algal-precipitate interactions. Final Report to the Canada Centre for Mining and Energy Technology, Energy, Mines, and Resources Canada DSS FILE NO.: 034SQ.23440-1-9011. https://zone.biblio.laurentian.ca/handle/10219/3069

Kalin, M., Romanin, B., & Mallory, G. (1995). Ecological engineering—A decommissioning technology. In B. J. Scheiner, T. D. Chatwin, H. El-Shall, S. K. Kawatra, & A. E. Torma (Eds.), *New remediation technology in the changing environmental arena* (pp. 71–76). Society for Mining, Metallurgy, and Exploration, Inc..

Kalin M., Meinrath, G., Smith, M. & Wheeler, W. (2011). Sustainable removal of Ra-226 and U from mine effluents: A review of field works in Northern Saskatchewan, Canada and Saxony, Germany. In *Uranium Mine Remediation Exchange Group*, UMREG, Selected papers 1995–2007, pp. 271-287. https://inis.iaea.org/search/search.aspx?orig_q=RN:43011875

Kamennaya, N., Ajo-Franklin, C., Northen, T., & Jansson, C. (2012). Cyanobacteria as biocatalysts for carbonate mineralization. *Minerals, 2*(4), 338–364. https://doi.org/10.3390/min2040338

Kaplan, D., Christiaen, D., & Arad, S. M. (1987). Chelating properties of extracellular polysaccharides from *Chlorella* spp. *Applied and Environmental Microbiology, 53*(12), 2953–2956. Retrieved from http://www.ncbi.nlm.nih.gov/pubmed/16347510%0Ahttp://www.pubmedcentral.nih.gov/articlerender.fcgi?artid=PMC204228

Kawahata, C., Yamamuro, M., & Shiraiwa, Y. (2013). Changes in alkaline band formation and calcification of corticated charophyte *Chara globularis. SpringerPlus, 2*(1), 85. https://doi.org/10.1186/2193-1801-2-85

Krienitz, L., Krienitz, D., Dadheech, P. K., Hübener, T., Kotut, K., Luo, W., Teubner, K., & Versfeld, W. D. (2016). Food algae for lesser flamingos: A stocktaking. *Hydrobiologia, 775*(1), 21–50. https://doi.org/10.1007/s10750-016-2706-x

Kufel, L., & Kufel, I. (2002). Chara beds acting as nutrient sinks in shallow lakes — A review. *Aquatic Botany, 72*(3–4), 249–260. https://doi.org/10.1016/S0304-3770(01)00204-2

Lacerda, L. D., Fernandez, M. A., Calazans, C. F., & Tanizaki, K. F. (1992). Bioavailability of heavy metals in sediments of two coastal lagoons in Rio de Janeiro, Brazil. *Hydrobiologia, 228*, 65–70. https://doi.org/10.1007/BF00006477

Lanzén, A., Simachew, A., Gessesse, A., Chmolowska, D., Jonassen, I., & Øvreås, L. (2013). Surprising prokaryotic and eukaryotic diversity, community structure and biogeography of Ethiopian soda lakes. *PLoS ONE, 8*(8). https://doi.org/10.1371/journal.pone.0072577

Leitch, A. R. (1991). Calcification of the Charophyte *Oosporangium*. In R. Riding (Ed.), *Calcareous Algae and Stromatolites* (pp. 204–216). https://doi.org/10.1007/978-3-642-52335-9_12

Li, Z., Zhang, Z., Jiang, W., Yu, M., Zhou, Y., Zhao, Y., & Chai, Z. (2008). Direct measurement of lanthanum uptake and distribution in internodal cells of *Chara. Plant Science, 174*(5), 496–501. https://doi.org/10.1016/j.plantsci.2008.01.013

Lobban, C. S., & Wynne, M. J. (1981). *The biology of seaweeds* (Vol. 17). Univ of California Press.

Mann, H., Fyfe, W. S., & Kerrich, R. (1988). The chemical content of algae and waters: Bioconcentration. *Toxicity Assessment, 3*(1), 1–16. https://doi.org/10.1002/tox.2540030103

Manusadžianas, L., Vitkus, R., Gylyt, B., Cimmperman, R., Džiugelis, M., Karitonas, R., & Sadauskas, K. (2020). Ecotoxicity responses of the macrophyte alga *Nitellopsis obtusa* and freshwater crustacean *Thamnocephalus platyurus* to 12 rare earth elements. *Sustainability, 12*(17), 7130. https://doi.org/10.3390/su12177130

Marquardt, M., & Schubert, H. (2009). Photosynthetic characterization of *Chara vulgaris* in bioremediation ponds. *Charophytes, 2*, 1–8. https://www.researchgate.net/publication/229066985

Marques, A. M., Bonet, R., Simon-Pujol, M. D., Fuste, M. C., & Congregado, F. (1990). Removal of uranium by an exopolysaccharide from *Pseudomonas* sp. *Applied Microbiology and Biotechnology, 34*(3), 429–431.

McConnaughey, T. (1991). Calcification in *Chara corallina* carbon dioxide hydroxylation generates protons for bicarbonate assimilation. *Limnology and Oceanography, 36*, 619–628. https://doi.org/10.4319/lo.1991.36.4.0619

McConnaughey, T. A., & Falk, R. H. (1991). Calcium-proton exchange during algal calcification. *Biological Bulletin, 180*, 185–195. https://doi.org/10.2307/1542440

McConnaughey, T. A., & Whelan, J. F. (1997). Calcification generates protons for nutrient and bicarbonate uptake. *Earth Science Reviews, 42*, 95–117. https://doi.org/10.1016/S0012-8252(96)00036-0

McQueen, N., Kelemen, P., Dipple, G., Renforth, P., & Wilcox, J. (2020). Ambient weathering of magnesium oxide for CO_2 removal from air. *Nature Communications, 11*(1), 1–10. https://doi.org/10.1038/s41467-020-16510-3

Meyers, R. A. (1977). Chemical desulfurization of coal. In *AICHE Symposium Series* (Vol. 73, 165, pp. 179-182).

Moheimani, N. R., Webb, J. P., & Borowitzka, M. A. (2012). Bioremediation and other potential applications of coccolithophorid algae: A review. *Algal Research, 1*(2), 120–133. https://doi.org/10.1016/j.algal.2012.06.002

Myers, V. B., Iverson, R. L., & Harriss, R. C. (1975). The effect of salinity and dissolved organic matter on surface charge characteristics of some euryhaline phytoplankton. *Journal of Experimental Marine Biology and Ecology, 17*(1), 59–68.

Neihof, R. A., & Loeb, G. I. (1972). The surface charge of particulate matter in seawater. *Limnology and Oceanography, 17*(1), 7–16.

Paasche, E. (1968). Biology and physiology of Coccolithophorids. *Annual Review of Microbiology, 22*(1), 71–86. https://doi.org/10.1146/annurev.mi.22.100168.000443

Parkhurst, D. L., & Appelo, C. A. J. (2013). Description of input and examples for PHREEQC version 3—A computer program for speciation, batch-reaction, one-dimensional transport, and inverse geochemical calculations. *U.S. Geological Survey Techniques and Methods,* Book 6, U.S. Geological Survey Techniques and Methods.

Pełechaty, M., Pukacz, A., Apolinarska, K., Pełechata, A., & Siepak, M. (2013). The significance of *Chara* vegetation in the precipitation of lacustrine calcium carbonate. *Sedimentology, 60,* 1017–1035. https://doi.org/10.1111/sed.12020

Pertl-Obermeyer, H., Lackner, P., Schulze, W. X., Hoepflinger, M. C., Hoeftberger, M., Foissner, I., & Obermeyer, G. (2018). Dissecting the subcellular membrane proteome reveals enrichment of H+ (co-) transporters and vesicle trafficking proteins in acidic zones of *Chara* internodal cells. *PLoS ONE, 13*(8), 1–28. https://doi.org/10.1371/journal.pone.0201480

Pukacz, A., Pełechaty, M., & Frankowski, M. (2014). Carbon dynamics in a hardwater lake: Effect of charophyte biomass on carbonate deposition. *Polish Journal of Ecology, 62*(4), 695–705. https://doi.org/10.3161/104.062.0413

Pukacz, A., Pełechaty, M., & Frankowski, M. (2016). Depth-dependence and monthly variability of charophyte biomass production: Consequences for the precipitation of calcium carbonate in a shallow Chara-lake. *Environmental Science and Pollution Research, 23*(22), 22433–22442. https://doi.org/10.1007/s11356-016-7420-8

Raven, J. A., & Beardall, J. (2003). Carbon acquisition mechanisms of algae: Carbon dioxide diffusion and carbon dioxide concentrating mechanisms. In A. W. D. Larkum, S. E. Douglas, & J. A. Raven (Eds.), *Photosynthesis in Algae* (pp. 225–244). https://doi.org/10.1007/978-94-007-1038-2_11

Raven, J. A., Smith, F. A., & Walker, N. A. (1986). Biomineralization in the Charophyceae sensu lato. In B. S. C. Leadbeater & R. Riding (Eds.), *Biomineralization in Lower Plants and Animals* (pp. 125–139). Clarendon.

Ray, S., Klenell, M., Choo, K.-S., Pedersén, M., & Snoeijs, P. (2003). Carbon acquisition mechanisms in *Chara tomentosa*. *Aquatic Botany, 76*(2), 141–154. https://doi.org/10.1016/S0304-3770(03)00035-4

Riding, R. (2000). Microbial carbonates: the geological record of calcified bacterial-algal mats and biofilms. *Sedimentology, 47*(suppl 1), 179–214. https://doi.org/10.1046/j.1365-3091.2000.00003.x

Romanin, B. (1994). A Mathematical Model for Biological Polishing. In *Boojum Technical Reports*. Retrieved from https://zone.biblio.laurentian.ca/handle/10219/2875

Sánchez, A. S., Nogueira, I. B. R., & Kalid, R. A. (2015). Uses of the reject brine from inland desalination for fish farming, *Spirulina* cultivation, and irrigation of forage shrub and crops. *Desalination, 364*, 96–107.

Sand-Jensen, K., Jensen, R. S., Gomes, M., Kristensen, E., Martinsen, K. T., Kragh, T., Baastrup-Spohr, L., & Borum, J. (2018). Photosynthesis and calcification of charophytes. *Aquatic Botany, 149*, 46–51. https://doi.org/10.1016/j.aquabot.2018.05.005

Schaivon, M., & Pilon-Smits, E. A. (2017). Selenium biofortification and phytoremediation phytotechnologies: A review. *Journal of Environmental Quality, 46*(1), 10–19. https://doi.org/10.2134/jeq2016.09.0342

Seckbach, J. (2007). Algae and cyanobacteria in extreme environments. In J. Seckbach (Ed.), *Cellular origin, life in extreme habitats and astrobiology, Vol 11.* Springer. ISBN 978-1-4020-6112-7.

Smith, F. A., & Walker, N. A. (1980). Photosynthesis by aquatic plants: Effects of unstirred layers in relation to assimilation of CO_2 and HCO^{3-} and to carbon isotopic discrimination. *New Phytologist, 86*, 245–259. https://doi.org/10.1111/j.1469-8137.1980.tb00785.x

Song, G. L., & Atrens, A. (1999). Corrosion mechanisms of magnesium alloys. *Advanced Engineering Materials, 1*(1), 11–33.

Sooksawat, N., Meetam, M., Kruatrachue, M., Pokethitiyook, P., & Nathalang, K. (2013). Phytoremediation potential of charophytes: Bioaccumulation and toxicity studies of cadmium, lead and zinc. *Journal of Environmental Sciences, 25*, 596–604. https://doi.org/10.1016/S1001-0742(12)60036-9

Sooksawat, N., Meetam, M., Kruatrachue, M., Pokethitiyook, P., & Inthorn, D. (2016). Equilibrium and kinetic studies on biosorption potential of charophyte biomass to remove heavy metals from synthetic metal solution and municipal wastewater. *Bioremediation Journal, 20*(3), 240–251.

Sooksawat, N., Meetam, M., Kruatrachue, M., Pokethitiyook, P., & Inthorn, D. (2017). Performance of packed bed column using *Chara aculeolata* biomass for removal of Pb and Cd ions from wastewater. *Journal of Environmental Science and Health, Part A, 52*(6), 539–546.

Spear, D. G., Barr, J. K., & Barr, C. E. (1969). Localization of hydrogen ion and chloride ion fluxes in *Nitella*. *The Journal of General Physiology, 54*, 397–413. https://www.ncbi.nlm.nih.gov/pmc/articles/PMC2225932/

Steinberg, C. E. W., Schäfer, H., Siedler, M., & Beisker, W. (1996). Taxonomic assessment of phytoplankton integrity by means of flow cytometry. *Archives of Toxicology, Supplement, 18*, 417–434.

Steinberg, C. E. W., Schäfer, H., Tittel, J., & Beisker, W. (1998). Phytoplankton composition and biomass spectra created by flow cytometry and zooplankton composition in mining lakes of different states of acidification. In W. Geller, H. Klapper, & W. Salomons (Eds.), *Acidic mining lakes. Environmental science*. Springer. https://doi.org/10.1007/978-3-642-71954-7_7

Sterritt, R. M., & Lester, J. N. (1979). The microbiological control of mine waste pollution. *Minerals and the Environment, 1*(2), 45–47. https://doi.org/10.1007/BF02010716

Stewart, K. D., & Mattox, K. R. (1975). Comparative cytology, evolution and classification of the green algae with some consideration of the origin of other organisms with chlorophylls a and b. *The Botanical Review, 41*(1), 104–135.

Stockner, J. G. (1988). Phototrophic picoplankton: An overview from marine and freshwater ecosystems. *Limnology and Oceanography, 33*(4), 765–775. https://doi.org/10.4319/lo.1988.33.4part2.0765

Stockner, J. C., & Antia, N. J. (1986). Algal picoplankton from marine and freshwater ecosystems: A multidisciplinary perspective. *Canadian Journal of Fisheries and Aquatic Sciences, 43*(12), 2472–2503.

Stumm, W., & Morgan, J. (1996). *Aquatic chemistry: Chemical equilibria and rates in natural waters*. Wiley, 3rd ed.

Sun, W., Xiao, E., Kalin, M., Krumens, V., Dong, Y., Ning, Z., Liu, T., Sun, M., Zhao, Y., Wu, S., Mao, J., & Xaio, T. (2015). Remediation of antimony-rich mine waters: Assessment of antimony removal and shifts in the microbial community of an onsite field-scale bioreactor. *Environmental Pollution, 215*, 213–222. https://doi.org/10.1016/j.envpol.2016.05.008

Van Raam, J. (2008). Bibliography of the Characeae. *Journal of the Indian Botanical Society*, 1–287.

Waisel, Y. (2012). *Biology of halophytes*. Elsevier.

Zhu, T., & Dittrich, M. (2016). Carbonate precipitation through microbial activities in natural environment, and their potential in biotechnology pp126: A review. *Frontiers in Bioengineering & Biotechnology, 20*, January 2016 | https://doi.org/10.3389/fbioe.2016.00004

Zimmermann, U., & Dainty, J. (Eds.). (1974). Membrane transport in plants. Springer, New York. https://link.springer.com/book doi:10.1007/978-3-642-65986-7. Print ISBN 978-3-642-65988-1. Online ISBN 978-3-642-65986-7.

Chapter 8
The Biofilm Generation Tool for the Reduction of Sulfate Oxidation

Margarete Kalin-Seidenfaden ⓘD

Abstract Weathering of rocks is enhanced by microbial activity. The same weathering process is encountered in the metal corrosion industry. MIC (Microbially Induced Corrosion) is a serious challenge to steel and other construction materials. Phosphate is known to serve as a corrosion inhibitor in metals. Applying this knowledge to mine wastes, a number of studies have added phosphate to produce an iron phosphate coating on the mineral surface. However, results have varied, and lacked longevity.

When a heap leach pile in Canada stopped working, Boojum thought this might be due to phosphate coatings on the ore surfaces. However, little phosphorus was found in SEM/EDX pictures of the heap leach rock. Only traces of phosphorus were found, but microbes were plentiful. Controlled experiments outdoors (in 75 L drums) were carried out with carbonaceous phosphate mining waste (CPMW). The effluents were monitored for about 2.7 years with nearly neutral effluent from the CPMW drums. Independent studies adding CPMW to mining wastes produced organic coatings on mineral surfaces. Lastly, the mechanisms were documented which led to the improved effluent. Acidophilic heterotrophic microbes overgrow and cover acidophilic chemo-autotrophs, forming a biofilm. To address biofilm longevity, the experimental rocks were stored under varying conditions for 11 years.

Keywords Oxidation reduction · Inhibition of weathering · Acidophilic heterotrophs · Chemolithotrophs · MIC (Microbially Induced Corrosion) · Encapsulation · Phosphate · Natural phosphate rock

M. Kalin-Seidenfaden (✉)
Boojum Research Ltd., Toronto, ON, Canada
e-mail: margarete.kalin@utoronto.ca

© The Author(s), under exclusive license to Springer Nature Switzerland AG 2022
M. Kalin-Seidenfaden, W. N. Wheeler (eds.), *Mine Wastes and Water,*
Ecological Engineering and Metals Extraction,
https://doi.org/10.1007/978-3-030-84651-0_8

8.1 History of Encapsulation Efforts

Working on different mine waste management sites, many researchers have recognized that ARUM (Acid Reduction Using Microbiology) and biological polishing tools have a difficult time when the pH of the drainage drops below 3. To prevent effluents from reaching pH 3, the weathering of waste rock and tailings must be slowed or stopped. Others have attempted to create coatings on the mineral surfaces, which would inhibit weathering rates by reducing oxygen access. Pioneers of this approach were the research group at the West Virginia Acid Mine Drainage Task Force (Meek, 1983, 1991; Renton & Stiller, 1988; Renton et al., 1988; Hart et al., 1990; Hart & Stiller, 1991). They added phosphate ore (natural phosphate rock; NPR) to coal waste piles at different dosages. This material is also used to fertilize acid soil since it contains both carbonates and phosphate. Evangelou (1995) postulated that, after the NPR dissolved, an iron-phosphate precipitate would coat the mineral surface, reducing oxygen access. The NPR additions produced a somewhat improved effluent, but the lowest dosage of NPR produced the best results, a high pH with alkalinity, in contrast to higher dosages where the pH was considerably lower, with metal acidity. Further studies to produce a coating continued (Mauric & Lottermoser, 2011; Harris & Lottermoser, 2016), with limited success. An extensive discussion on micro-encapsulation is given by Sahoo et al. (2013). Unfortunately, the encapsulation efforts along with many other attempts using soaps, cyanide, antibiotics, etc., failed to produce an economic solution to halt oxygen and water entering the waste piles.

These efforts performed relatively well for a few years, and in some cases, for one to two decades. In many cases, however, problems arose within the first decade of implementation, caused mostly by ecological, geochemical and microbiological processes which altered the initial conditions. The common limitations reflect the view that the problem is purely an inorganic chemical, engineering and hydrological problem (Table 8.1).

Table 8.1 presents some of the most common, currently used oxidation-inhibiting strategies for sulfidic mining wastes

Oxygen Control Technique Examples	Common Limitations
Dry soil composite covers (capillary breaks)	Imperfect (cracking, erosion), bioturbation colonizing by deep rooted vegetation,
Compost or bio-solids covers	Finite organic supply
Wet covers (dam impoundments)	Dam vulnerability and wave turbulences and water shortages possible
Lake or pit backfilling	Lack of sites & regulatory constraints, ground water ingress
Desulfurized tailing slimes cover	Requires iron sulfide separation with separated sub-aqueous impoundment and sand backfilling - limited applicability
Co-deposition (rock and tailings)	Limited opportunity, timing of waste generation, distribution problems

Although air and water are essential drivers of the contaminant generating process, it is not the only relevant one. The methods implemented in the past to control oxidation of mining wastes have limitations. Omitting or disregarding ecological processes in the engineering of the mine waste management areas, severely constrains the longevity of any engineered solution. Understanding and integrating ecological and biochemical processes involved in contaminant generation and the natural biological strategies available to reduce or inhibit sulfide oxidation in mine wastes are critical, cost effective, more energy-efficient, and may lead to sustainable restoration of mine waste management areas (Kalin, 2009).

An interesting discovery came in the form of an unusual occurrence. An article in the Northern Miner reported that a heap leach dump in British Columbia had ceased to function, generating pH neutral effluent (Scott, 1991). For Boojum, the questions were: Why had the heap leach stopped working? Was it due to microbial or chemical inhibition? Could this be a phosphate coating which had formed on the mineral surface?

Rocks from this heap leach operation were obtained to examine the surface for the presence of microbes and/or iron phosphate coatings (Fig. 8.1). When the surfaces were inspected with SEM (Scanning Electron Microscopy) and EDX (X-ray diffraction), very little phosphorus was noted, but a considerable number of microbial colonies were evident on the rock surface. With these results, the West Virginia experiments needed to be repeated. The challenge was how to obtain waste rocks and tailings in sufficient quantities to give statistically significant results, and to set up a controlled, outdoor experiment at the Boojum Research (Boojum) facility.

Fig. 8.1 Rock from a heap leach that ceased generating acid. The rock was examined with SEM and EDX for iron phosphate or other coatings, which may have prevented leaching

8.2   Repeating the Experimentation: The Big Surprise

Microbially-Induced Corrosion (MIC) is responsible for the deterioration of steel and other industrial materials (Little & Wagner, 1992; Wakefield & Jones, 1998; Beech et al., 2000; Kip & Van Veen, 2015). To reduce corrosion, steel and iron are now routinely phosphated (Schweitzer, 1988). The microbial processes that apply to MIC in steel and other industrial materials also apply to stone and rock and thus to mine waste rock and tailings. The most common growth form for microbes is the biofilm. Characklis & Marshall (1990) published the first book on biofilms, addressing the growing awareness among scientists and engineers that microbial biofilms play an essential role in engineering processes.

In the 1990s, when Boojum first started to tackle this issue, neither the dynamics of biofilm development, nor microbial biochemical activities had been elucidated. Biofilms and their behavior are now well-documented (Sand & Gehrke, 2006; Gorbushina & Broughton, 2009; Flemming & Wingender, 2010; Zhang et al., 2016). Bacterial biofilms were suspected to be the reason the heap leach stopped working, but this hypothesis needed to be tested.

If microbes were producing biofilms, covering mineral surface and blocking sulfate oxidation, the process should be the same in both tailings and waste rock. Further, if the reduction in weathering was due to microbial activity, then the microbial weathering process should be the same in both the field and laboratory and should work in different mining wastes. The key was feeding heterotrophic microbes to make suffocating biofilms, and starving acidophilic chemo-autotroph biofilms which produce acidic drainage.

The first experiment was carried out with three tonnes of differently-aged sulfidic waste rock (i.e., various states of oxidation). Rocks were shipped from a Quebec zinc mine to Toronto where the experiment was set up outdoors at Boojum's facility (Fig. 8.2). The waste rock was divided among 12, 75 L plastic drums. Half of the drums contained 8 L of carbonaceous phosphate mining waste (CPMW) added to the top of the drum, distributed throughout the entire drum or held in by mosquito netting placed in the middle of the drum. The latter application of netting simulated the application of CPMW to lifts of a growing waste rock pile. The other half of the drums were left without CPMW, as controls. The CPMW material used, C-48 (Nutrien Co., Aurora Mine), was a coarse by-product remaining after finer material is separated for use as fertilizer. It contained 30% calcium and 6.8% phosphate.

After 8.6 years outdoors, including winters, and a long, dry storage, and repeated outdoor exposure, the drums were dismantled and the surface area of pyrite estimated (Kalin et al., 1998). About 50% of the 8 L of the CPMW remained in the drums, unreacted, some slightly coated with iron. Selected rocks from the drum experiment were subjected to SEM/EDX (Scanning Electron Microscopy/Energy Dispersive X-ray Spectroscopy). The surfaces of rocks with added phosphate wastes showed only traces of phosphorus, while microbes were relatively abundant (Fyson et al., 1995; Kalin & Harris, 2005).

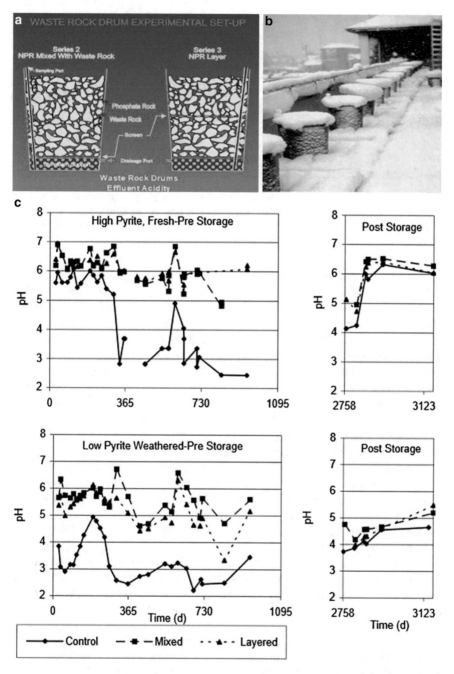

Fig. 8.2 Waste rock drum experiment on the balcony of the Boojum Research office. (**a**) Schematic showing two ways CPMW was added to waste rock in the drums. Each drum contained 8 L of CPMW waste added to the top of the drum, distributed throughout the entire drum or held in by mosquito netting placed in the middle of the drum. With these distributions of CPMW to lifts of a

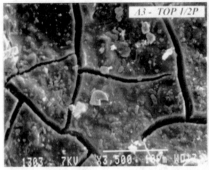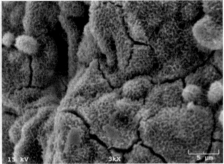

Fig. 8.3 SEM photographs of the surface of sulfidic waste rock. The left side shows the rock surface after 2.7 years of exposure to the elements and CPMW. On the right side, the SEM photograph of the surface coating on sulfidic waste rock after the second outdoor exposure, 11 years since the start of the experiments (from Kalin et al., 2010)

Once the effluents reached a neutral pH, investigations of the rock surfaces were initiated (Fig. 8.3). The formation of an organic coating was confirmed (Ueshima et al., 2003, 2004). The longevity of the organic coating was tested by dry-storing the rocks for several years. Thereafter, the waste rocks were re-exposed to the elements outdoors without further additions of CPMW. Repeated recycling of rainwater provided ideal conditions for acid generation. As expected, the pH dropped initially, but then increased with reduced acidity (Fig. 8.2c). This outdoor exposure continued for 2.7 years. The rocks were dry-stored again for further investigations of the mineral surface. The coating appeared to be long-lived, persisting for 4.5 years indoors, then for another 2 years outdoors, and finally further drying stored in a basement. SEM photographs, after 11 years from the start of the experiment, revealed that the organic coatings were still present (Kalin et al., 2010).

The results suggested that CPMW additions catalyze a fundamental change on the surface of the waste rock. Organic coatings form over the sulfides. Whether on waste rock or in tailings of different composition, CPMW wastes should behave in a similar manner, and react with sulfide minerals. If this assumption is correct, and heterotrophic microbes were responsible for the coating, then the CPMW waste ingredients were key to reducing oxidation.

These experimental results, addressing the reduction of sulfate oxidation, are probably the first which span more than a decade (Fig. 8.3). Of interest, though, is the biofilm discovered on an excavated Roman nail (reported in Kip & Van Veen, 2015).

Fig. 8.2 (continued) waste rock piles are simulated. (**b**) Photograph of the drums outdoors, in the winter. (**c**) The effluent pH from the drums (control, mixed, and layered waste rock). Effluent was collected periodically for 2.7 years or 1095 days. The drums were then stored indoors for about 4.3 years (up to day 2756). The stored rocks were again placed back outdoors to test longevity of the biofilms without any further CPMW added. Acidic effluent was initially generated, but soon the pH rose again. Details in Kalin & Harris (2005). On dismantling of the drums, about 50% of the CPMW had remained in the drums. (Photograph by Boojum Research)

This nail was uncorroded after several thousand years of burial, suggesting that microbial biofilms can persist and protect surfaces for millennia.

Given that biofilms operate on the sulfide mineral surface, heterotrophic biofilms should work on sulfide minerals whether they are on waste rock or tailings. To address the fundamental nature of the findings, CPMW needed to be added to different sulfidic materials. A few mining companies agreed to finance shipments of phosphate waste rock from North Carolina to various mining sites in quantities large enough to set up field experiments (several railcars and trucks). The phosphate mine shipped the wastes, washed and screened, to Ontario (Elliot Lake, Sudbury, and South Bay), as well as to Nova Scotia (Cape Breton). The Newfoundland experiments used CPMW from a closed phosphate fertilizer plant, which had received ore from the same mine in North Carolina.

Field plots were set up by plowing CPMW wastes into the upper 25 cm of different tailings. It was hypothesized that phosphate added to these tailings would produce an iron phosphate precipitate, generating a hardpan within the tailings, reducing infiltration, while feeding heterotrophic microbes to cover the tailings granules with biofilms (see Fig. 9.8). Tests were conducted on fresh pyrrhotite tailings, on a concentrate spill, and on waste coal piles. The plots were left to the elements for 3 to 4 years (Table 8.2).

The results are summarized for a uranium mine, pyrrhotite tailings, a polymetallic concentrate spill, and the rock drainage from Quebec in Table 8.2 and in Kalin et al. (2009). In the upper part of Table 8.2, the length of time in years is given for the plots in the field (from set up to sampling time), the time of dry storage, and the laboratory monitoring times, exchanging the supernatant intermittently. In the lower part of the table, the CPMW application rates, number of monitoring events where pH, E_h and electrical conductivity (measurement pairs) were obtained, the hydraulic conductivity, and the site-specific sulfide range in % are given.

Table 8.2 Exposure and storage times and the type of monitoring of the CPMW additions to different wastes

	Uranium	Pyrrhotite	Polymetallic	Waste rock
Length of time (years)				
Exposure (outdoors)	3.75	3.25	3.18	2.7
Storage indoors	6.5	5.5	n.a	n.a
Monitoring of Effluents weight/vol(1:5 w/v)	1.83	1.83	2.87	2.7
Selected Samples				
CPMW particles mixture (ϕ) 0.01-0.25 m) (ϕ) 4 to <0.04 mm Application rate in the field	30 kg. m^{-2}		1:4 (w/w)	115 kg.t^{-1}
Number of measurements E_h-pH pairs	7/7	7/7	8/8	58/115
General characteristics				
Hydraulic conductivity (cm.s^{-1})	10^{-5}	10^{-8}	10^{-5}	10^{-0}
Site specific sulfide % of wastes	2	85	6-8	4-15

n.a. - not applicable; (a) #Eh-pH pairs

After 3 or 4 years of exposure in the field, plot markings were impossible to decipher and many were destroyed. Boojum therefore collected samples randomly on the field plots and returned them to the laboratory, where the material was added to beakers and covered with water. The supernatant (water above the sediment) in the beakers was exchanged 7 to 8 times over a period of 1.8 to 2.8 years.

Visually, the CPMW plots could not be recognized in the field samples. We monitored the supernatant (Table 8.3), exchanging the water and mixing with a vortex shaker. The pH changed initially indicating some acid generation, but further acid generation was not evident. Many samples were processed, but due to the destroyed field plot makings, only a few samples showed an effect (Kalin et al., 2003). The supernatant samples, with the elevated pH and a corresponding control, were analyzed for elemental concentrations. Slight hints of inhibition might be interpreted from the slightly higher calcium concentrations, the increases in pH, and the decreases in the electrical conductivity, but there were no significant changes. Plot markings need to be kept prominent for the next experimental iteration.

An assessment of the oxidation rate of iron sulfide in tailings with and without CPMW was carried out. The samples originated from tailings in an advanced state of sulfide oxidation. The oxidation rates should have been in equilibrium with

Table 8.3 Chemistry of slurries of the sampled field plots and exposed to oxygen in the laboratory

Elements (mg.L^{-1})	Uranium		Pyrrhotite		Polymetallic		Sulfidic Waste rock	
	Control	CPMW	Control	CPMW	Control	CPMW	Control	CPMW
	N=1	N=1	N=1	N=1	N=2	N=2	N=3	N=6
Al	50	<0.005	870	120	89	5	3.98	0.42
Ca	560	630	500	490	485	510	41	122
Cu	0.59	0.001	0.68	17	86	0.07	14	4.13
Fe	18	0.01	43	0.1	1053	0.02	3.77	0.32
P	0.03	0.05	0.22	6.9	0.16	0.04	0.20	0.13
S	630	510	4460	1060	3020	500	126	123
Zn	0.98	<0.005	9.3	5.3	2085	22	78	22
pH	2.6	6.7	3.1	3.8	3.0	5.0	4.0	6.1
Cond. (µS.cm^{-1})	3410	1682	7180	4030	6725	1690	n.a.	n.a.
E$_h$ (mV)	734	584	758	661	784	467	n.a.	n.a.
Acidity (mg.L^{-1})	656	39	6715.4	1090	5544	87	257	76

Table 8.4 Derived oxidation rates for tailings compared to those calculated by Williamson and Rimstidt (1994)

		E_h (V)		pH	
		min	max	min	max
Model Williamson and Rimstidt (1994)		0.805	0.917	1.57	2.53
Uranium tailings	Control	0.734	0.804	2.42	2.92
	CPMW	0.566	0.750	6.06	7.17
Pyrrhotite tailings	Control	0.577	0.758	2.98	3.50
	CPMW	0.635	0.772	3.34	3.84
Polymetallic tailings	Control	0.445	0.585	2.19	3.22
	CPMW	0.198	0.523	3.92	5.38
Sulfidic	Control	0.467	0.814	2.37	5.41
Waste rock	CPMW	0.41	0.72	4.01	6.29

respect to E_h. The rate laws (r) described by Williamson and Rimstidt (1994) were used to calculate oxidation rates. Using E_h/pH as a basis for calculating log r in samples exposed to field conditions (under conditions promoting oxidation), empirically integrates all the processes controlling oxidation. These are likely a combination of both chemical and geo-microbiological processes (Table 8.4).

8.2.1 Composition and Dissolution

The composition of CPMW is as critical as its leaching characteristics to an understanding of how CPMW interacts with biofilms and the microbial processes leading to the inhibition or reduction of sulfide oxidation (Kalin & Wheeler, 2011; Kalin et al., 2012).

The elemental composition of unprocessed natural phosphate rock has been described by (Kalin et al., 1998). Waste phosphate material contains about 8 to 9 % P, 25 to 27% Ca, 3% Fe and 0.5 to 0.1% Na, Al, Mg and K, respectively. Solubility tests were carried out with 0.5 g of CPMW in 125 mL acid-leached poly-propylene bottles with dilute aqueous H_2SO_4, adjusted to pHs 3, 5 and 6.5. Aliquots were tested after 10, 30, 60 and 120 min of reaction time. Dissolved orthophosphate was determined using Hatch Phosphover reagent and a DREL 2000 spectrophotometer. The concentration of orthophosphate in the pH 5 and 6.5 bottles was below 1 mg.L^{-1}, but in pH 3 bottles, 2 mg.L^{-1} was released after 60 min.

Stronger leach tests were carried out in 0.1 N H_2SO_4, simulating possible conditions in corrosion pits on mineral surfaces and in tailings pore water. Ten grams of CPMW were exposed to 125 mL of acid solution, stirred briefly, allowed to settle, and then decanted. New acid was added and the procedure repeated for a total of 8 decant cycles collecting a total of 1L of leachate. The CPMW was stirred briefly after the sulphuric acid addition and pH was measured until no further changes were noted. Samples of the solution were submitted to ICP (Inductively Coupled Plasma

Spectroscopy) and a mass balance (% weight solid element released to the sulphuric acid) was determined (Kalin et al., 2009).

The next experiment dissolved CPMW with 0.1 N sulfuric acid (pH 1.9) to determine not only phosphate, but other elements released from the material under stirred and non-stirred conditions. The findings support the low release of phosphate, even in very strong acid. Only 54% of phosphate was recovered in the supernatant, under both conditions. In addition, 26% of the calcium, 75% of the potassium and 100% of the magnesium were released.

Estimating the quantity of dissolved orthophosphate released from CPMW is associated with several uncertainties. When the drums from the original experiment (Kalin & Harris, 2005) were dismantled about half of the added CPMW had remained unreacted or slightly covered with iron. This suggests that the dissolution rate for CPMW is low. Since other elements are also released, one or more of the components could be the critical factor in the production of the organic coatings.

8.3 Rocks on the Move: Independent Investigations of CPMW

Research results should be reproducible. To follow up on Boojum's research findings discussed above, waste rocks from Quebec and CPMW were shipped to Dr. R Smart (University of South Australia) for an AMIRA (Australian Mineral Industries Research Association) project, investigating the organic coating on the Quebec rocks and from Red Dog mine (a Canadian zinc mine) waste rocks. The electron microscope investigation identified secondary mineral enclosures in the organic coating on the Quebec rocks, and an organic coating on the Red Dog waste rocks (AMIRA, 2017; Kalin et al., 2010). A list of researchers, reports, organizations etc. are detailed in (Table 8.5).

From the very beginning when the drum experiment was set up, Boojum was certain that the mining industry would be very interested in the results, but each time results were presented, more information was required. The last request was that without the names of the microbes at work, the results were useless. The taxonomy, however, is irrelevant, of importance is their activity and the longevity of the biofilms. Biofilms are composed of a community of microbes, which might be different at each location. Most important, though, are the metabolic results, namely oxygen consumption and heterotrophic microbial biofilm growth. If heterotrophic biofilm growth could be documented, along with reductions in oxidation rates and improved effluent, a direct connection would be made.

Table 8.5 A list of researchers and organizations that have contributed to investigations on different pyritic wastes

Author	#Reports	Organisation	Waste Type
Kalin, M.	31	Boojum	All types
Smith, M.P.	10	Boojum	Coal, Nova Scotia
Fyson, A.	10	Boojum	Rock, Quebec
Wheeler, W.N.	6	Boojum	All types
Paulo, C.	4	Boojum	Rock, Quebec
Fortin, D.	3	U of Ottawa	Rock, Quebec
Meinrath, G.	2	RER	All types
Ueshima, M.	1	U of Ottawa	Rock, Quebec
Bellenberg, S.	1	U of Duisburg Essen	Coal, Germany
Sand. W.	1	U of Duisburg Essen	Coal, Germany
Smart, R.	1	U of South Australia	Rock, Yukon
Sleep, B.	1	U of Toronto	Rock, Quebec
Ferris, G.	1	U of Toronto	Rock, Quebec
Harris, B.	1	U of McGill	Rock, Quebec
Werker, A.	1	U of Waterloo	All types
Totals:	**15**	**8**	**8[a]**

[a]Uranium, pyrrhotite and zinc tailings others sulfidic rock

8.4   Eureka – The Microbial Groups Are Identified

CPMW provides a combination of nutrients essential to the growth of heterotrophs on the mineral surface. No link between phosphate content, biofilm growth, and dormancy could be found, but carbonates in the CPMW seem to influence heterotrophic biofilm development. To determine which microbial groups might be active, Boojum supported a PhD student at the Biofilm Centre of the University of Essen (Germany) to carry out a controlled experiment on German lignite coal (Bellenberg et al., 2013).

Laboratory columns were set up with lignite coal and ground pyrite. In the absence of CPMW (labelled NPR in Fig. 8.4), bioleaching started after day 45 with the release of ferric iron, increases in redox potential and electrical conductivity, and decreasing pH values. The addition of CPMW inhibited bioleaching, resulting in an improved effluent quality. No change of the microbial community composition occurred without addition of CPMW, while its addition strongly proliferated neutrophilic heterotrophs. The CPMW application led concurrently to a reduction in the number of acidophilic iron oxidizers. Four weeks after the CPMW addition, the columns were dominated by neutrophilic heterotrophs (90%), while without the CPMW, iron-oxidizing acidophiles accounted for 99% of the microbial community. A photographic record of the clear solutions from the CPMW-treated columns along with the usual parameters measured is shown in Fig. 8.4 (Bellenberg et al., 2013). Additional information can be found in Kalin et al. (2015, 2018).

The development of the microbial community composition in percent and pyritic lignite colonization 10 weeks after application of NPR.

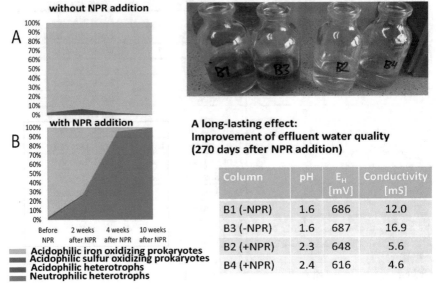

A long-lasting effect:
**Improvement of effluent water quality
(270 days after NPR addition)**

Column	pH	E_H [mV]	Conductivity [mS]
B1 (-NPR)	1.6	686	12.0
B3 (-NPR)	1.6	687	16.9
B2 (+NPR)	2.3	648	5.6
B4 (+NPR)	2.4	616	4.6

Fig. 8.4 Results of the CPMW coal column experiments in Germany. (**a**) The development of microbial populations on the mineral surface with and without CPMW. Oxidizing microbes (green color) produce acid drainage, whilst neutrophilic heterotrophs (blue color) replace the oxidizing microbes. (**b**) Photograph of effluents from column experiments after 213 days. Jars B1 and B3 had no CPMW added, whilst B2 and B4 contained coal and CPMW. Effluents from the latter are clear, free of iron, in contrast to B1 and B3 columns without CPMW. Unfortunately, there is a red box behind B4. (**c**) The table shows the change in pH, E_h, and conductivity in effluents from columns with and without CPMW (NPR; Bellenberg et al., 2013)

Sustainable mine waste and water management should integrate the use of carbonates when constructing a waste rock pile, and when initiating a hardpan in tailings is appropriate. The improved effluent of the first experiments summarized in Kalin and Harris (2005) and the presented results of the field and laboratory work should be sufficient evidence that contaminated acid drainage is not the price to be paid for metals.

CPMW has been shown to alter the mineral surface on waste rock, ground coal, and tailings. The formation of neutrophilic, heterotrophic biofilms, produced in the presence of CPMW, has been well documented here. The longevity of these biofilms has been tested - they have been shown to be actively protecting mineral surfaces after at least 11 years. Biofilms appear to protect mineral and metal surfaces, and this protection can last for thousands of years (Kip & Van Veen, 2015). These pieces of evidence are compelling, yet Boojum has not found a client willing to follow up with further testing. Perhaps Boojum has found the answer to the biggest environmental challenge in mining? More evidence can be found in the next chapter (Chap. 9) which describes several case studies using CPMW.

References

AMIRA. (2017). Alternative Treatments for ARD Control. Retrieved from http://www.amira.com. au/web/site.asp?section=projects&page=projectdetails&ProjectLink=2861&Source_ID=1

Beech, I., Bergel, A., Mollica, A., Flemming, H., Scotto, V., & Sand, W. (2000). Microbially influenced corrosion of industrial materials. *Recent Advances in the Study of Bio-corrosion. Task 2:* Brite-Euram III Thematic Network ERB BRRT-CT98-5084 Biocorrosion 00-02 (pp. 177–190).

Bellenberg, S., Kalin, M., & Sand, W. (2013). Microbial community composition on lignite before and after the addition of phosphate mining wastes. *20th International Bio-Hydrometallurgy Symposium.* Antofagasta. Chile. Abstract #978-0-0001-488-7-600-SB. Retrieved from: https://books.google.com/books/about/20th_International_Biohydrometallurgy_Sy.html?id=kERRjwEACAAJ

Characklis, W. G., & Marshall, K. C. (Eds.). (1990). *Biofilms* (Vol. 5). Wiley-Interscience.

Evangelou, V. P. (1995). *Pyrite oxidation and its control.* New York: CRC Press. pp 293 ISBN 9780849347320 - CAT# 4732.

Fyson, A., Kalin, M., & Smith, M. P. (1995). Reduction of acidity in effluent from pyrite waste rock using natural phosphate rock. pp. 270–277. In *Proceedings of the 27th Annual Meeting of Canadian Mineral Processors.* Ottawa. Ontario. January 17–19.

Flemming, H. C., & Wingender, J. (2010). The biofilm matrix. *Nature Reviews, 8,* 623–633. https://doi.org/10.1146/annurev.micro.091208.073349

Gorbushina, A. A., & Broughton, W. J. (2009). Microbiology of the atmosphere-rock interface: How biological interactions and physical stresses modulate a sophisticated microbial ecosystem. *Annual Review of Microbiology., 63,* 431–450. https://doi.org/10.1146/annurev.micro.091208.073349

Harris, D., & Lottermoser, B. (2016). Phosphate stabilization of polymetallic mine wastes. *Mineralogical Magazine, 70*(1), 1–13. Retrieved 3 Nov. 2019. https://doi.org/10.1180/0026461067010309

Hart, W., Stiller, A., Rymer, T., Renton, J., Skousen, J., Sencindiver, J., & Samuel, D. (1990). The use of phosphate refuse as a potential AMD ameliorant. In *Proceedings of the Mining and Reclamation Conference and Exhibition.* West Virginia University Publications Service. No. 1. pp. 43–49.

Hart, W., & Stiller, A. H. (1991). Application of phosphate refuse to coal mine refuse for amelioration of acid mine drainage. *Second International Conference on Abatement of Acidic Drainage.* Montreal. Canada: MEND. Ottawa pp. 173–190.

Kalin, M. (2004). Slow progress in controlling acid mine drainage (AMD): A perspective and a new approach. *Pekiana. Staatliches Museum für Naturkunde Görlitz, 3,* 101–112.

Kalin, M. (2009). Ecological perspectives in restoring mine waste management areas. *Proceedings of the 48th Conference of Metallurgists (COM 2009).* Sudbury. Ontario. Canada.

Kalin, M., & Harris, B. (2005). Chemical precipitation within pyritic waste rock. *Hydrometallurgy., 78*(3–4), 213–229. http://www.boojumresearch.com/content/EcolEng-J13.pdf

Kalin, M., & Wheeler, W. N. (2011). A review of the role of phosphate mining waste: A chemical or biological reagent for AMD prevention. *West Virginia Mine Drainage Task Force Symposium Papers.* Morgantown, W. V. March 29–30 (pp. 1–12). Retrieved from https://wvmdtaskforce.com/past-symposium-papers/2011-symposium-papers/

Kalin, M., Smith, M. P., & Fyson, A. (1998). The role of phosphate in applied biotechnology in mine waste management: reduction in AMD from pyritic waste rock. *Proceedings of the International Symposium. the Metallurgical Society of the Canadian Institute for Mining. 'Waste Processing and Recycling.'* Calgary, Alberta. August 16-19. pp. 15-29.

Kalin, M., Fyson, A., Smith, M. P., & Werker, A. (2003). Tailings surface cover development through integration of reactive phosphate and organic matter. In G. Spiers. P. Beckett. & H. Conroy (Eds.). *Proceedings of "Mining and Environment III."* (pdf #80). Sudbury: Laurentian University. Retrieved from https://zone.biblio.laurentian.ca/handle/10219/3019

Kalin, M., Ferris, G., & Paulo, C. (2009). Reducing sulphide oxidation in pyritic mining wastes – phosphate mining wastes stimulate biofilm formation on the mineral surface. In Proceedings of the 8th ICARD meeting, "Securing the future." Skelleftea, Sweden. http://www.proceedings-stfandicard-2009.com/

Kalin, M., Paulo, C., & Sleep, B. (2010). Proactive prevention of acid generation: Reduction/inhibition of sulfide oxidation. In C. Wolkersdorfer & A. Freund (Eds.), *Proceedings. Mine Water and Innovative Thinking* (International Mine Water Association Symposium) (pp. 479–482). Cape Breton University Press.

Kalin, M., Paulo, C., Smart, R., & Wheeler, W. (2012). Phosphate mining waste reduces microbial oxidation of sulfidic minerals: A proposed mechanism. *International Conference on Acid Rock Drainage (ICARD)*. (paper #0272). Ottawa: MEND Publications.

Kalin, M., Paulo, C., Sudbury, M. P., & Wheeler, W. N. (2015). Reducing sulfide oxidation in mining wastes by recognizing the geomicrobial role of phosphate in mining wastes—A long journey 1991–2014. *Journal of the American Society of Mining and Reclamation., 4*(2), 102–121. https://www.asmr.us/Portals/0/Documents/Journal/Volume-4-Issue-2/Kalin-CA.pdf

Kalin, M., Wheeler, W. N., & Bellenberg, S. (2018) Acid Rock Drainage or not - oxidative vs. reductive biofilms—A microbial question? *Minerals* 2018. *8*(5). 199; Retrieved from: https://doi.org/10.3390/min8050199

Kip, N., & Van Veen, J. A. (2015). The dual role of microbes in corrosion. *ISME Journal. 9*(3). 542–551. Retrieved from: https://doi.org/10.1038/ismej.2014.169

Little, B., & Wagner, P. (1992). Recent advances in the study of microbiologically influenced corrosion. In *MRS proceedings* (Vol. 294, p. 343). Cambridge University Press.

Mauric, A., & Lottermoser, B. G. (2011). Phosphate amendment of metalliferous waste rocks. Century Pb–Zn mine. Australia: Laboratory and field trials. *Applied Geochemistry, 26*(1), 45–56. https://doi.org/10.1016/j.apgeochem.2010.11.002

Meek, F.A. (1983). Research into the use of apatite rock for acidic drainage prevention. *Fifth Annual West Virginia Mine Drainage Task Force Symposium*. Morgantown. WV https://wvm-dtaskforce.files.wordpress.com/2015/12/84-meek.pdf

Meek, A. F. (1991). Assessment of acid prevention techniques employed at the Island Creek Mining. In *Twelfth Annual West Virginia Surface Mine Drainage Task Force Symposium* April 3–4. Morgantown West Virginia (pp. 21–28). Retrieved from https://wvmdtaskforce.files.word-press.com/2015/12/91-meek.pdf

Renton, J. J., & Stiller, A. H. (1988). Use of phosphate materials as ameliorants for acid mine drainage. Volume 1. The use of rock phosphate (apatite) for the amelioration of acid mine drainage from the mining of coal. *Final report* (No. PB--93-216448/XAB). West Virginia Univ., Morgantown. WV.

Renton, J. J., Stiller, A. H., & Rymer, T. E. (1988) The use of phosphate materials as ameliorants for acid mine drainage. *Conference on Mine Drainage and Surface Mine Reclamation* (Vol. 1. pp. 67-75). United States Bureau of Mines Information Circular IC 9183.

Sahoo, P. K., Kim, K., Equeenuddin, S. M., & Powell, M. A. (2013). *Current Approaches for Mitigating Acid Mine Drainage BT - Reviews of Environmental Contamination and Toxicology Volume 226* (D. M. Whitacre (ed.); pp. 1–32). Springer New York. https://doi.org/10.1007/978-1-4614-6898-1_1

Sand, W., & Gehrke, T. (2006). Extracellular polymeric substances mediate bioleaching/biocorrosion via interfacial processes involving iron (II) ions and acidophilic bacteria. *Research in Microbiology, 157*, 49–56.

Schweitzer, P. A. (1988). *Corrosion and corrosion protection handbook* (2nd ed.). CRC Press.

Scott, D. (1991). Gibraltar's dump leaching—An insight into acid effluent control. *The Northern Miner., 77*(30), 1–7.

Ueshima, M., Kalin, M., & Fortin, D. (2003). Microbial effects of natural phosphate rock (CPMW) addition to mining wastes. *Joint Conference of the 9th Billings Land Reclamation Symposium and the 20th Annual Meeting of the American Society of Mining and Reclamation* (pp. 1294–1303).

Ueshima, M., Fortin, D., & Kalin, M. (2004). Development of iron-phosphate biofilms on pyritic mine waste rock surfaces previously treated with natural phosphate rocks. *Geomicrobiology Journal., 21*(5), 313–323.

Wakefield, R. D., & Jones, M. S. (1998). An introduction to stone colonizing micro-organisms and biodeterioration of building stone. *Quarterly Journal of Engineering Geology and Hydrogeology., 31*(4), 301–313.

Williamson, M. A., & Rimstidt, J. D. (1994). The kinetics and electrochemical rate-determining step of aqueous pyrite oxidation. *Geochimica et Cosmochimica Acta., 58*(24), 5443–5454. https://doi.org/10.1016/0016-7037(94)90241-0

Zhang, R., Bellenberg, S., Neu, T. R., Sand, W., & Vera, M. (2016). The biofilm lifestyle of acidophilic metal/sulfur-oxidizing microorganisms. In P. H. Rampelotto (Ed.), *Biotechnology of extremophiles: Advances and challenges* (pp. 177–213). https://doi.org/10.1007/978-3-319-13521-2_6

Chapter 9
R&D Field Applications

Margarete Kalin-Seidenfaden ⓘ

Abstract This chapter describes several major projects undertaken by Boojum Research since 1982 in Canada, Brazil, and Germany. These projects ran between 5 and 16 years. The biological polishing tool reduced the concentrations of arsenic and nickel in a 5 million cubic meter pit lake and also the concentrations of zinc and iron from the effluent of a gloryhole using a series of polishing ponds. In Germany, a constructed wetland with rooted vascular plants was expected to reduce the concentrations of radionuclides, arsenic and iron from the effluent of a uranium mine. The removal was inefficient. The settling pond of the constructed wetland was used as a pilot system, by suspending curtains and introducing *Chara*, the multitasking alga, to sequester contaminants.

Using Carbonaceous Phosphate Mining Waste (CPMW), Acid Reduction Using Microbiology (ARUM) and biological polishing, a complete decommissioning concept was developed for an operating zinc mine in Quebec. Similarly, all tools were partially scaled up at a copper-zinc mine in northern Ontario. Finally, a coarse coal pile in Nova Scotia was treated with CPMW and compared to other piles treated with limestone, and at a different Nova Scotian site, the tools were used to treat effluents from waste coal tailings.

Keywords Decommissioning · Acid mine drainage · Alkaline drainage · Mine waste management · Sulfide oxidation · Carbonaceous phosphate mining wastes · Ecological engineering

M. Kalin-Seidenfaden (✉)
Boojum Research Ltd., Toronto, ON, Canada
e-mail: margarete.kalin@utoronto.ca

© The Author(s), under exclusive license to Springer Nature Switzerland AG 2022
M. Kalin-Seidenfaden, W. N. Wheeler (eds.), *Mine Wastes and Water,
Ecological Engineering and Metals Extraction*,
https://doi.org/10.1007/978-3-030-84651-0_9

9.1 Arsenic and Nickel Removal from a Pit Lake – Saskatchewan

A mined-out uranium pit in northern Saskatchewan was force-flooded from an adjacent fishing lake. Once filled, the lake would hold 5 million m^3, with the only influxes coming from precipitation. With environmental approval, the barrier between the pit lake and an adjacent fishing lake would be removed. The arsenic and nickel concentrations of 0.22 mg.L^{-1} and 0.26 mg.L^{-1}, respectively, were low, but exceeded regulatory limits. Before the lake and the pit water could be combined, the arsenic and nickel concentrations had to be lowered. Reductions in these low concentrations were not only difficult to achieve with chemical treatment, but also expensive. If biological polishing could be used to reduce the concentration further, chemical treatment costs could be avoided. The company decided to fund a detailed investigation using biological polishing to remove the nickel and arsenic (Fig. 9.1).

Four sets of sedimentation traps were placed at various depths to collect suspended solids. The traps suspended at 3 m (close to the extinction of light) collected particulates and algae, which were subjected to SIMS microscopy (Secondary-Ion Mass Spectroscopy). This method examines the surface of materials to a depth of 1 to 2 nanometers. From the data generated by SIMS, a schematic was constructed (Fig. 9.2).

The force-flooding produced high concentrations of total suspended solids (TSS) in the first 2 years. The first summer after flooding, the particulates were mainly inorganic clay which gave way in the second year to an algal bloom of *Dictyosphaerium* sp. (Fig. 9.3a, b). These algae form colonies of cells embedded in a dense extracellular polysaccharide (EPS) sheath. The dense sheath around the algae was also a 'magnet' for inorganic particulates. The algae were acting like

Fig. 9.1 Sediment traps used to collect and analyze particulates. (**a**) The traps on the left were installed at depths between 3 and 45 m. The traps closest to the boat had a fine layer of clay which was not visible on any of the following traps. (**b**) Photograph of the 3 m depth traps one year after flooding, showing the high quantity of suspended solids (Boojum Research, 1994; Cao & Kalin, 1999; Kalin et al., 2001). (Photographs by Boojum Research)

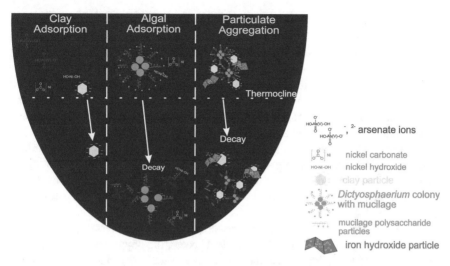

Fig. 9.2 Schematic of particle aggregation based on SIMS microscopy, depicting algae as flocculating agents. The green circles and spirals are algal cells and EPS molecules, reddish brown shapes symbolize iron hydroxide, yellowish shapes are clay particles, pink shapes are nickel hydroxide, and violet crosses represent arsenic.

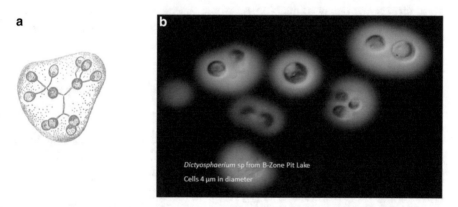

Fig. 9.3 (**a**) Drawing of a colony of *Dictyosphaerium*, showing the 'connected' nature of the algal cells. (**b**) Photograph of the sheath, likely an EPS compound (Boojum Research, 1994, 1997). (Photograph by Boojum Research)

living flocculants, removing inorganic particulates, arsenic and nickel from the water column.

If the algae and associated particulates were relegating the contaminants to the sediments, the nickel and arsenic removal from the water column should balance with the arsenic and nickel found in the sediments. To produce this mass balance, bottom sediments were retrieved at about 35 m depth in several locations on the pit lake bottom with an Eckman grab sampler. The top 5 cm were retrieved from the sampler, dried and submitted for elemental analysis, pooling the samples. The concentrations of relevant elements in the sediment were compared to the sum of the elemental concentrations determined from the sedimentation traps over eight years. The results were close for the two contaminants, As and Ni (Table 9.1).

The elemental concentrations in the sediment traps compared well with the same elemental concentrations in the top 5 cm of the bottom sediment (Table 9.1). Exceptions were aluminum and iron, with higher concentrations in the sediment than found in the traps. Aluminum discrepancies can be explained by the presence of a layer of clay which was placed at the bottom of the pit to cover hazardous materials before filling. The higher iron concentrations could be explained by grabs that caught muskeg sediment that would have increased iron concentrations. Given these numbers, though, the mass balance survived the scrutiny of the company engineers.

The next step was to support and possibly increase the growth of the phytoplankton in the pit lake. In the first couple of years, the phytoplankton population density was relatively low, and a nutrient limitation was suspected. Using a simple Redfield ratio (Falkowski, 2000), it was estimated that an addition of 720 kg of calcium nitrate would probably serve to maintain or increase the biomass, thereby increasing the polishing capacity (Boojum Research Ltd., 1997; Kalin et al., 2002; Kalin & Wheeler, 2013). One to two fertilizer applications over time would suffice, since growth and decay of the biomass would recycle the nutrients within the pit lake. It is unknown if this recommendation was implemented, as the project ended in 2003.

Table 9.1 Mass balance for elements found in particulates collected in sedimentation traps between 1992 and 1999 compared to sediment grab samples

Year	Sediment rate	Sedimentation per year g/m$_2$	As	Ni	Fe	Al	PO$_4$
1992	28.7	10,472	1.5	1.7	184	161	24
1993	28.7	10,472	3.8	4.7	99	77	22
1994	11.6	4227	7.9	4.5	60	38	8
1995	13	4758	12.8	3.4	62	30	9
1996	3.5	1285	2.7	0.9	26	18	2
1997	2.4	876	6.7	1.1	41	14	3
1998	2.6	931	4	0.9	44	17	2
1999	2	737	2.7	0.4	38	10	2
Totals		33,757	42	18	555	366	73
In Sediment 1999 Depth 0–7 cm			30	17	566	619	57

9.2   Zinc Removal from Circum-Neutral Gloryhole Effluent – Newfoundland

Boojum Research was retained in 1988 to evaluate the applicability of using eco-logical engineering measures at a very complex mining system in Newfoundland. The old, polymetallic mine operated from the 1920s to 1984, and used the historic gloryhole method of mining. The ore was taken along the ore veins, dumped on rail cars into a gloryhole, and then transported through a haulage tunnel to the mill. Many orebodies were connected to the haulage tunnel, which, at the time of decom-missioning, was referred to as the drainage tunnel. The major contaminant was zinc, which would precipitate with oxidized iron at neutral pH. At decommissioning, all workings were hydraulically connected and nearly all discharged to the Oriental East Pit (OEP), and thence to the Buchans River. Adjacent to the OEP (pH 5.8 to 7.0) is the Oriental West Pit (OWP; pH range between 2.7 and 3.4) which has no direct outflow. Both pit lakes did not stratify, but froze during the winter. The com-bined effluents exiting from the OEP contained between 20 and 25 mg.L^{-1} zinc throughout the year.

One of the goals of the project was to lower the concentrations of zinc that flowed into the Buchans River from the OEP. Since there was a large meadow downstream from the OEP, Boojum decided to use biological polishing to treat the effluent. A series of 6 small ponds were constructed and filled with cut brush as substrate for periphyton growth (Fig. 9.4).

These pilot, biological polishing ponds were shallow, and froze during the win-ter. Later, these 6 pilot ponds were expanded to cover the entire meadow. The scaled-up ponds worked only in the summer dropping the zinc concentration to 5 mg.L^{-1} or below. Winter removal rates were essentially absent. Winter results were attributed to ice which covered the OEP, preventing oxidation and particle formation. Since the iron could not oxidize, iron and zinc did not precipitate, hence no zinc removal. Periphyton on the cut brush collected the iron/zinc particles in the summer reducing zinc and iron concentrations (Fig. 9.5).

The two scaled-up ponds closest to the OEP, ponds 10 and 14, would have to cleaned out periodically, as they would fill up with precipitate quite rapidly (see Fig. 9.6). But, before the ponds could fill with precipitate, the project came to an abrupt halt, when the mine manager passed away suddenly in 1995. Shortly after, the responsibility for the site changed hands and a consultancy took over manage-ment. Boojum obtained system monitoring data from the Newfoundland govern-ment in 2007, when Boojum representatives visited the site. Upon reviewing the site, they were stunned, as the alder cuttings, serving as substrates for the periphyton had been removed. The locals, who had worked with Boojum in the past reported that the consultancy had requested that all brush be cleaned out of the pools. Shortly thereafter, a truckload of sugar arrived with instructions to the local caretaker to deposit it into the OEP. Sugar dissolves and leaves with the water! It was clear from the government monitoring data, that the ponds had ceased working sometime after

Fig. 9.4 (**a**) Martin Smith, a Boojum researcher, working on one of the pilot ponds. Note brush in the pond, used as a substrate for periphyton growth. (**b**) A schematic showing the pilot-scale ponds in yellow, and the scale-up in blue. The series of 6 yellow circles extending through pools 14 and 15 were the pilot ponds which served as design criteria for the scale up. First to be scaled up were pools 10–13 (in lighter blue), followed later by pools 14–17 (darker blue). Yellow arrows denote direction of flow. The three yellow circles in the lower left were ARUM test cells treating the drainage from the waste rock pile. Organics were added to a completely rust-covered sphagnum bog. (**c**) Aerial photograph of the OEP and polishing ponds in the meadow below the OEP. All photographs by Boojum Research.

the consultancy took over management. Ecological systems need to be monitored, and managed with knowledge of their function.

The effluents from the OEP were not the only challenge. Zinc ore concentrate (55% zinc) had spilled onto a water-saturated muskeg area below the mill. The acidity values ranged between 1,000 and 10,000 mg.L^{-1} CaCO$_3$. Here, Boojum recommended pilot tests with CPMW from a shutdown phosphate plant available on the island. It was anticipated that an application might lead to a hardpan forming a drain which would precipitate some of the zinc along the way.

The mine manager had established several 5 L open bottom buckets with different addition ratios between (1:4 and 1:20) over the muskeg area. The buckets were sent to the Boojum laboratory after a short exposure of 10 days and after a 3 year exposure in the muskeg area. In the laboratory, the same routine was used as

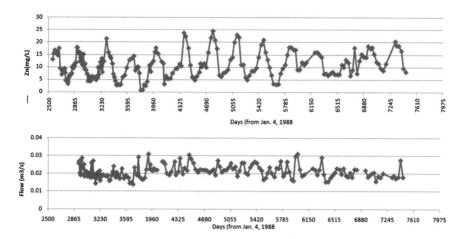

Fig. 9.5 Flow and zinc removal from biological polishing ponds at Buchans, Newfoundland over a period of 13 years. (Data provided by the Newfoundland Ministry of Environment)

Fig. 9.6 Two photographs of periphyton heavily-caked with oxidized iron at each weir within the pond system and at its edges. (Photograph taken in 2009 by Boojum Research)

described in Chap. 8 with tailings. Supernatants were produced and the samples were kept under oxidizing conditions. The experiment was terminated after 1200 days (Kalin, 2004). The most important observations were that, regardless of the mixing ratios, the pH and acidity in the supernatants remained constant or improved compared to controls (Fig. 9.7a). An application of CPMW would produce lower acidity and zinc concentrations being discharged into the river. Shortly before his death, the manager had the local CPMW spread onto the muskeg area with the concentrate spill (Fig. 9.8). The project is described in detail in Kalin (2009).

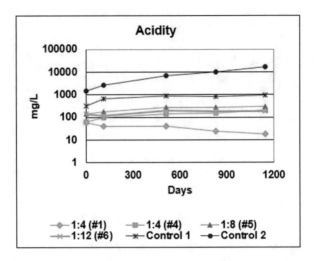

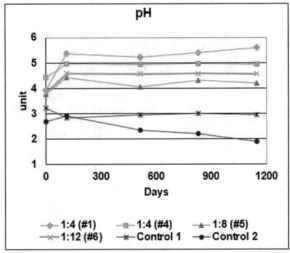

Fig. 9.7 Acidity and pH of bucket samples from the concentrate spill. The acidity (**a**) and pH (**b**) are shown for different ratios of concentrate soil mixtures (w:w, soil:CPMW) over time. Red lines are two different controls, i.e., no additions of CPMW. These data show a decline (control 2) in pH after a slight initial increase over the time, or almost no change in pH for control 1. Acidity of controls remained fairly constant after an initial increase. The best performance was noted in sample number one with ratio 1:4, indicating that, once reacted, no further acidity was produced

9.3   Limestone and CPMW Application to Coal Waste Piles – Nova Scotia

At a coal mining and processing facility in Nova Scotia, metallurgical and lower grade coal was produced. Four relatively large waste piles were set up as test piles, where management was addressing various methods of integrating a layer of

Fig. 9.8 Boojum Research personnel monitoring the grass cover on the tailings concentrate spill in 2009. Photograph by Boojum Research

limestone to reduce acid mine drainage. A fifth pile was added, to which CPMW was supplemented in the same configuration as the other 4 piles (Fig. 9.9a). All five piles were heavily instrumented with lysimeters to collect drainage. The work was carried out by an engineering consultancy.

Four piles produced drainage, but the fifth pile, with CPMW, produced no or very little drainage. Instead, the layer of CPMW produced large erosion channels along the sides of the pile. The same consultancy that had set up the test piles was hired by Boojum to perform an autopsy of all test piles (Fig. 9.9b). An extensive hard pan had formed within the CPMW pile, but was not found in any of the piles where limestone was added. The hardpan was extensively documented, and may have been self-sealing (Fig. 9.9b; Baechler, 1997).

Concurrently, Boojum set up laboratory columns which were monitored periodically by adding water and collecting effluent (Fig. 9.9c). The two columns with the yellow or brown stripes produced less and less drainage, eventually plugging. Boojum monitored the collected drainage from the columns for several years, but failed to obtain funding to analyze the drainage and complete the data interpretation. Both the field piles and laboratory columns demonstrated that CPMW could form hardpans, preventing the penetration of precipitation.

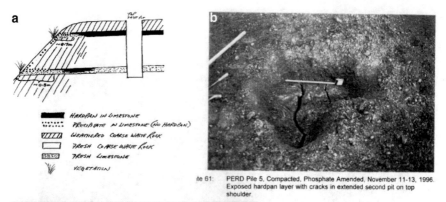

Plate 61:   PERD Pile 5. Compacted, Phosphate Amended, November 11-13, 1996.
             Exposed hardpan layer with cracks in extended second pit on top
             shoulder.

Fig. 9.9 Experiments using CPMW on coal piles in Nova Scotia. (**a**) Schematic for the application of both CPMW and of differently-compacted limestone, testing construction of coarse coal waste piles. (**b**) An autopsy of all piles showed the pile containing CPMW produced a hardpan, which reduced infiltration, preventing acidic effluents. Note hardpan layer at bottom of excavated pit with cracks which were described as self-healing, i.e., closed at bottom of crack. (**c**) Laboratory columns with hardpan formation seen in rust-colored layers set up in the Boojum laboratory. Column photograph by Boojum Research, coal pile photograph and schematic by Fred Baechler

9.4 Tailings Hardpan Development Ontario

In general, a hardpan in tailings is desirable, as it prevents or slows penetration of precipitation, and thus, the oxidation of the sulfides. Ideally, this hardpan should be generated in the tailings just before the facility is shut down. Boojum experimented using CPMW in both fresh pyrrhotite tailings and aged uranium tailings.

At a mine site in Sudbury, Ontario. Boojum set up experiments with CPMW plowed into the fresh tailings (Fig. 9.10a). At this site, straw and grass seed were crimped into the tailings. At the aged tailings site (a uranium mine in Elliot Lake, Ontario; Fig. 9.10b), instead of straw, horse manure was crimped with the grass seed. CPMW was tilled into the tailings with a handheld plow. These measures

Fig. 9.10 (**a**) Fresh pyrrhotite tailings in N. Ontario. CPMW was plowed into the tailings in the same fashion as lime would be integrated in tailings for revegetation. Straw was crimped into the surface and the tailings were seeded to control erosion. (**b**) The plots on abandoned uranium tailings. The control plots are shown in the foreground, while the plots with the highest CPMW and organic applications are in the dark area with a Boojum employee. All plots were seeded with manure and grass seed. (Photographs by Boojum Research)

(grass seed, horse manure, and straw) were used to prevent erosion. Samples from the plots were collected after a little more than 3 years in the field.

The application of CPMW was expected to form a hardpan to reduce infiltration of rain and snowmelt. After 3 years, the markers delineating the plots had been lost. Since it was manually impossible to dig up the plots looking for the hardpan, an EM39 conductivity meter was used. This instrument performs a geophysical survey with fixed frequency electromagnetic (EM) profiling techniques employing a Geonics EM39 instrument. The EM survey was supposed to find differences in apparent conductivity in plots with and without CPMW additions. The instrument provided measurements of both the quad-phase (conductivity) and in-phase (magnetic susceptibility) components within two distinct depth ranges, simultaneously. However, the apparent conductivity, EC_a, of the plots, which should have indicated the hardpan, failed because the surrounding conductivity was out of range for the instrument, i.e., the control plots and the surroundings were highly conductive. Perhaps if a backhoe had been available, the hardpan would have been found. Different EMs (EM 31- DL Em31 and Em34-3 instruments were very successful on the mine site locating boreholes, shafts and adits where acid drainage production occurred (Kalin & Pawlowski, 1994; Hutchinson & Barta, 2000).

While a hardpan could not be confirmed in the fresh and aged tailings, its presence in the Nova Scotian coal piles and in Newfoundland concentrate spill suggests that this approach was worth trying on a larger scale. The application of CPMW has to be simple, with easily-available equipment. The dosage of CPMW was estimated by the mine operator based on the same cost of hauling and distributing limestone. It is hoped that commercial applications of CPMW to actual, large-scale waste rock piles and tailings will reduce the production of weathering products leading to an improved quality of drainage, but these tests have yet to be performed.

9.5 Decommissioning Concepts Applying ARUM, CPMW and Biological Polishing – Quebec

Both tools, ARUM and biological polishing, were sufficiently understood at the time Boojum received a contract to develop a decommissioning plan for a zinc mine in Quebec. However, at that time, there was only limited evidence to suggest that microbes and phosphate wastes might be playing important roles slowing pyrite oxidation. The literature on phosphate and acid mine drainage was growing. Evangelou (1995), Georgopoulou et al. (1996), and Chen et al. (1997) carried out intensive studies on pyrite in coal, postulating that direct adsorption of phosphate molecules onto iron atoms on the pyrite surface leads to the subsequent elimination of electron transfer between pyrite and oxidizing agents. Boojum thought that applications of carbonaceous phosphate mine waste (CPMW) to mine waste rock and tailings might be beneficial in the reduction of sulfide oxidation. Thus, CPMW application became the cornerstone of the decommissioning scenario.

For Boojum, this mine waste management area in Quebec was the first opportunity to present an environmental management system for an operating mine (Boojum Research, 1992). All Boojum's ecological tools were needed, but, most importantly, tools for reducing sulfide oxidation. Although the waste rock pile was just about starting the second lift, acid drainage was already destroying the muskeg (Fig. 9.11). The photograph shows the brown dead muskeg where drainage took its toll. The green shrubs mark the future path of the ARD drainage. In the distance, a light brown line can be seen on the horizon. This was a new ditch which would collect future drainage to the chemical treatment plant. A thorough hydrological reconnaissance prior to locating the waste rock might have prevented or lessened the impact.

Cost associated with conventional decommissioning options for a zinc mine in Quebec were estimated at the start of the 1990s. Operation estimated that decommissioning costs would be from $10 to $50 million for the tailings, and $5 to $15 million for the waste rock pile. The mine's owner engaged the services of Boojum Research in 1990 to identify a less costly decommissioning scenario. Ideally, this scenario would not include the operation of a conventional treatment plant, which would be operated in perpetuity.

A plan by the mine's owner to neutralize the acid mine drainage produced during operation would require approximately 5.7 million tonnes of lime over 285 years and generate 30 million m^3 of sludge. The costs and sludge production estimates were provided by BP Selco. These outrageous numbers argued for a more sustainable and less costly alternative. An ecological engineering decommissioning

Fig. 9.11 Aerial photograph of an operating zinc mine in Quebec. Iron oxidation is visible as brown stain. The light brown line at the horizon is a ditch that had been dug to intercept the drainage leaving the property of the mine waste management area. (Photograph by Boojum Research)

concept was developed for the tailings and waste rock management areas, based on an assessment of an existing water quality monitoring program.

The decommissioning scenario Boojum suggested included adding a layer of granular CPMW to each waste rock lift. Ditches would be filled with haybales, initially to reduce the iron loading to the treatment plant. In time, these ditch treatment cells would be covered with floating, living islands, like those described in Chap. 6. In this proposal, Boojum would build ARUM cells to treat ARD in the perimeter ditch. Effluents would first encounter a cell containing CPMW to precipitate oxidized iron. The effluents would then enter several ARUM cells to remove acidity and metals. Finally, the 'cleaned' effluents would be directed to the finished pit. A schematic of the proposed system is shown in Fig. 9.12.

To test the CPMW application to waste rock, three tonnes of variously-sized rocks were shipped from the mine to the Boojum facility in Toronto. Rocks were distributed into 12, 55-gallon drums. CPMW was distributed in various ways to the drums. The drums sat outside for about 3 years, as natural precipitation drained through them. Details of this experiment are presented in Chap. 8.

Another experiment was set up on site in Quebec. This system was designed to test applications of CPMW, ARUM and biological polishing to the mine's effluents. The pilot treatment system is shown in Fig. 9.13. Here, effluents would enter a tank with CPMW, removing much of the oxidized iron. The water would then be processed first in a biological polishing tank, followed by an ARUM tank, and then finally a 2nd biological polishing tank. This system was built and treatment was initiated when the mine was sold, and the project terminated.

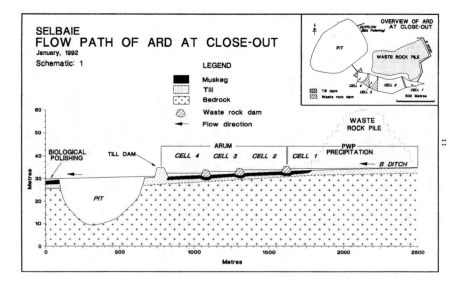

Fig. 9.12 Schematic showing the flow path of ARD and the proposed ARUM cells to treat it

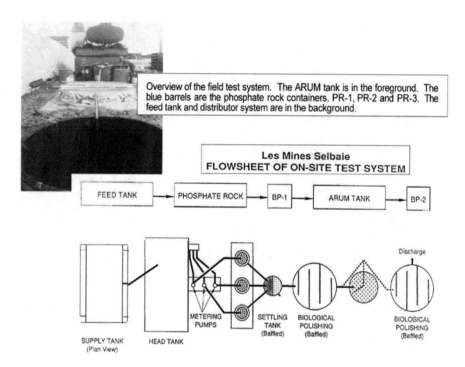

Fig. 9.13 Overview of pilot test system using CPMW, ARUM and biological polishing to treat waste rock ARD

The other major source of contaminated effluents was the tailings. In the decommissioning scenario developed for the mine, fresh tailings would be plowed with CPMW and seeded with grass. Boojum was given a completed section of tailings pond, together with a spigot line for experimentation.

Here, the transport of water through the tailings can be reduced with the formation of a hardpan within 0.5 m depth of the tailings surface (Fig. 9.14). The tailings above the hardpan are tilled with organics and grass seeds, as shown in the schematic shown in Fig. 9.14. Heterotrophic microbes living in the organics would remove oxygen from the upper strata of tailings. The combination of low oxygen and hardpan would restrict water and air access to the tailings, reducing the production of AMD.

The first experiment on fresh tailings involved adding vegetation seeds to the spigot line. Then, as the tailings were laid down, the seeds would sprout and cover the tailings with vegetation. However, this result did not occur. Tests carried out with this option failed, as growth only occurred in cracks of the drying tailings (Fig. 9.15a).

Field trials were carried out on fresh tailings using erosion control mats. A plot 70 m long by 8 m wide was established in June 1990 over a moisture gradient

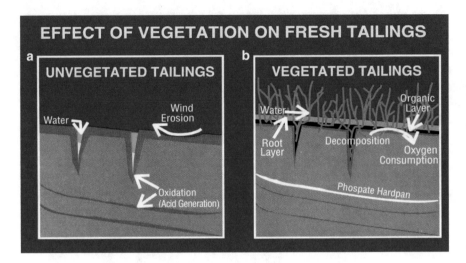

Fig. 9.14 (**a**) Schematic showing the concept of unvegetated (barren) tailings as they dry. Water and air enter drying cracks oxidizing pyrite in the tailings on the left. (**b**) Concept of tailings in which a layer of CPMW is added to fresh tailings, but buried to react in the vadose zone, as exemplified in section 9.6. The CPMW (phosphate hardpan) would initiate iron precipitation and, in part, neutralize the acid. Most important, though, would be the presence of heterotrophic microbes which would consume oxygen in the vadose zone

which existed along the tailings slope (wet, moist, and dry). Verdyol™ strips (4.0 × 3.75 m) were used to blanket the tailings. They were further divided into areas with different arrays of seeds and CPMW. Water penetrating the tailings should result in development of a low permeability stratum (iron precipitate hardpan) beneath the vegetation layer. This will provide suitable conditions for heterotrophic (oxygen-consuming) bacteria in the root zone of the grass layer, conventionally established (Kalin et al. 1993; Fig. 9.15b). Tailings, shortly after discharge, are not yet totally solid, requiring researchers like Martin Smith to constantly move (Fig. 9.15c).

A reduction in the rate at which tailings oxidize might be achieved if a population of oxygen-consuming bacteria (heterotrophs) could be maintained in the root region of a vegetation cover. Heterotrophs growing in the root zone would consume the oxygen in the water entering the tailings. Such bacteria have much higher potential growth rates than chemo-autotrophs and other pyrite-oxidizing bacteria. Boojum then asked the question: are there any heterotrophic, oxygen-consuming microbes in fresh tailings? Hence the entire mill circuit was tested for their presence and viability. Water samples at various stations throughout the mill were plated onto commercial agar dishes, and the viable colonies were counted (Table 9.2). The results were clear, tailings have viable heterotrophic microbe populations.

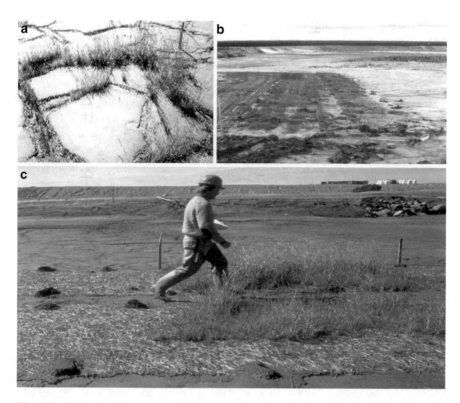

Fig. 9.15 Field experiments to grow grass and vegetation on fresh tailings. (**a**) Vegetation seeds were added to the fresh tailings spigot line. Results were not satisfactory, as vegetation only grew in the cracks. (**b**) The more satisfying efforts with straw mats and CPMW. (**c**) Boojum researcher, Martin Smith, monitoring the grass plots. (Photographs by Boojum Research)

Table 9.2 Bacterial colonies per milliliter of tailings supernatant at various positions along the mill circuit

Sample	Heterotrophic bacterial colonies	pH
Sagmill before addition	23	6.7
Sagmill after addition	550	7
Ballmill discharge	0	7.7
Thickener underflow	0	11.5
Thickener overflow	0	12.5
Barren tailings-surface	<1	n.d.
Barren tailings-sub-surface	6	n.d.
Tailings beach	1,404,000	8.1
Tailings seepage ditch 1	40,500	2.1
Tailings seepage ditch 2	10,800	4.8
Old Pond	9	2.2

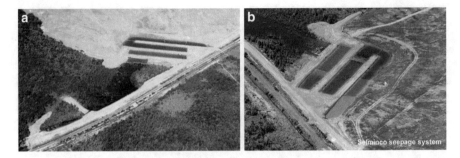

Fig. 9.16 A CPMW, ARUM, and biological polishing treatment system established on a coal tailings dump in Nova Scotia. (**a**) System shortly after installation. Iron oxidation was occurring throughout the ditch system. (**b**) System after installation of curtains and CPMW dikes were constructed with the coarse gravel to foster iron oxidation in the first cells only. ARUM cells are shown as clear of iron. (Photographs taken by Fred Baechler)

9.6   CPMW and ARUM in a Coal, Tailings Dump in Nova Scotia

A coal, tailings dump in Nova Scotia produced acidic effluent, which was impacting the local vegetation (Fig. 9.16). Boojum was tasked with decommissioning the effluent stream. The concept consisted of ditches which were to pretreat the unoxidized iron in the effluent with CPMW. The oxidized iron and other contaminants would then be treated downstream to remove acidity and metals using ARUM (Kalin 1993). Finally, as a last step, biological polishing would remove remaining contaminants. Fig. 9.16a shows the system in the first year after installation. At this point, iron was being oxidized throughout the ditch system, and fouling hay bales installed for ARUM. Installation of curtains in the first ditch (Fig. 9.16b), enhanced the oxidation process, allowing only oxidized iron and contaminants to enter ditches downstream. A test area with ARUM (and floating cattails) was planned once the system was working. The effectiveness was evident in the successively clearer water entering the proposed ARUM area (Fig. 9.16b).

9.7   Biological Polishing of ^{226}Ra, Iron and Arsenic – Germany

With the fall of the Berlin Wall, Wismut GmbH became responsible for the remediation of the East German uranium mining district in the Erzgebirge. Wismut had experimented with the classical wetland treatment for the removal of iron, arsenic, uranium and radium. They used plants rooted in wetland sediment which overwintered and regrew above-ground biomass every year. These aquatic plants are referred to in German as helophytes. The wetland was set up in an old swimming pool. One side housed a regular wetland with plants rooted in sediment. Iron oxidation was accomplished using cascade 'waterfall' on the left side of Fig. 9.17 shown as a green bar. The open water side of the pool was used to collect the iron precipitate. The results were not very satisfactory, and Boojum was ask to implement some of its ecologically engineered tools.

With iron precipitation in the cascade, arsenic and ^{226}Ra were partially co-precipitated and removed from the circumneutral water. At the recommendation of Boojum, mats or curtains were suspended in the open water part of the pool to grow periphyton and biofilms which were to polish the arsenic, iron and ^{226}Ra. Floating vegetation islands were tested as a means to supply dissolved organic carbon for biofilm development (Figs. 9.18 and 9.19).

Both the biofilm curtains and the above-ground plants were added to the pool at roughly the same time and grew for a maximum of 6 months. The biofilm scrapings were more effective at contaminant removal than the above-ground vegetation parts (Table 9.3). However, the root systems of the wetland vegetation were not compared to the biofilm scrapings. If they had been analysed, the differences might have been less. One potential drawback of the wetland approach was the potential for food chain contamination, which would not occur with biofilms and periphyton, as the latter would, ideally, become biomineralized. Given the radioactivity of the accumulating sediments, they were to be handled as low-level radioactive sludge and treated as any other sludge from a chemical treatment plant.

The results from the pilot project looked promising so the system was scaled up (Fig. 9.20a). Shortly after the scale-up, the management of the system was turned

Fig. 9.17 Wismut-designed mine water treatment system with rooted plants receiving water from the right side for treatment. On the right side, in the open water, Boojum recommended installing curtains for biofilm formation and floating cattail islands to address fine iron particles. In the foreground some cattail islands grew well, but others were plagued leaf lice. The light blue buckets were installed to provide a flow-through system for *Chara*. Growth data were used to derive design criteria for scale-up

Fig. 9.18 Curtains added to Wismut treatment system to support periphyton and microbial biofilms. (**a**) Closeup of one of the curtains, showing periphyton with precipitated iron. (**b**) Placement of the curtains in the Wismut treatment system, surrounding floating islands with cattail seedlings

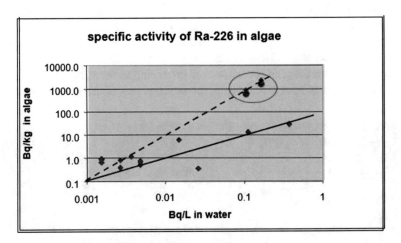

Fig. 9.19 Graph showing the specific activity of ^{226}Ra removal by *Chara* at different radium concentrations. The *Chara* removal line (dotted) is compared to removal rates by other periphyton (solid) from the treatment system. The lines suggest that *Chara* can remove low levels of ^{226}Ra, and it has a higher affinity for radium than other periphyton, even though there are only a few data points to support these conclusions. After 6 months in the containers, *Chara* thalli, starting from an initial concentration of < 1 Bq.kg^{-1}, had a specific activity of 8700 Bq.kg^{-1}, suggesting that *Chara* at high ^{226}Ra effluent concentrations showed a concentration factor greater than 1000

over to a consultancy, which discontinued Boojum's participation. After Boojum left the project, Prof. Schubert from the University of Rostock, an expert in *Chara* growth, was asked to determine the health of the *Chara* growing in the ponds. He compared the photosynthesis of *C. vulgaris* from an unpolluted reference site (Fig. 9.20b) and from the bioremediation pond (Fig. 9.20c).

Table 9.3 Table compares the two treatment systems, wetland and biofilm. Here, ^{226}Ra, iron, and arsenic removal are compared. In the upper box, biofilm/periphyton scrapings from the right-hand treatment system (including mats and floating vegetation structures) were compared to above-ground vegetation from the wetland treatment system

Contaminant removed by structures or helophytes			
Structures or helophytes	Fe mg/kg dry	As mg/kg dry	Ra-226 Bq/kg dry
Structures designed to remove contaminants			
Coconut matting	146,000	19,500	21,000
AQUA-mats[a]	129,000	17,900	22,000
Floating mat – *Juncus*	136,000	10,400	27,000
Floating mat – *Typha*	124,000	16,400	34,000
Floting mat – *Phragmites*	142,000	13,400	35,000
Contaminant concentrations in above-ground biomass of helophytes			
Phragmites communis	128–193	3–38	0.05–0.15
Juncus sp.	103–385	3–15	0.05–0.72
Typha latifolia	167–230	7–39	0.17–0.23
Iris pseudacorus	161–340	4–14	0.05–0.82

Contaminants removed by biofilms on mats and by emergent, above-ground plant parts (Wisutec, 2002)
Aquatic vegetation is labeled as helophytes

Fig. 9.20 Bioremediation ponds in Pöhla (Boojum and Wismut, GmbH), Germany (**a**). Scale-up of the pilot test system at Pöhla (**b**). Non-encrusted *C. vulgaris* from the unpolluted reference site near the bioremediation ponds. (**c**) Heavily encrusted *C. vulgaris* from the bioremediation pond, scale bar = 1cm. (Photographs: H. Schubert)

Even though the charophytes from the bioremediation ponds were heavily encrusted, they maintained photosynthetic activity. However, respiration rates of non-encrusted *C. vulgaris* were higher than those of encrusted individuals. After removal of the crust, rinsed specimens revealed intact thalli which showed similar respiration rates to non-encrusted *Chara* (Marquardt & Schubert, 2009). Several years later, Boojum was informed that the system had been totally shut down. This was quite disappointing, as so much time and effort had been invested, with excellent support from Wismut. However, Herbst et al. (2019) started to look at utilizing algae for a mine effluent in the Mansfeld region in Germany.

9.8   Decommissioning with ARUM, CPMW, & Biological Polishing – Ontario

During the operation of a copper/zinc mine in northern Ontario, one million tonnes of tailings were deposited on a peninsula adjacent to a trophy fishing lake, which formed the headwaters of the English River. The tailings contained 45% pyrite and 5% pyrrhotite. A chemical treatment plant installation was not feasible since there was no secured space for the generated sludge. Based on calculations, given the pyrite and pyrrhotite content, the site tailings were expected to generate acid for thousands of years (Kalin et al., 1992). For these two reasons, Boojum was contracted to develop a decommissioning plan for the site. With an agreement between the mine's operator and the Ontario regulatory agencies, the mine and surroundings were declared an R&D site in 1986 with the provision that no discharges reach the trophy fisheries lake. Ownership was returned to the crown, and funding for decommissioning was secured for 16 years. An overview of all of the tools and processes tested on the site is provided in Kalin (2003).

Within the mine waste and water management area, lay a 1 million cubic meter lake. The mine, when operated, generated drainage and underground seepages, which were diverted into the lake. Upon Boojum's arrival, the lake was already at pH 4. Boojum decided to use this lake as a large treatment pond, and experimented with a number of tools. Boojum's efforts were focused mainly on stabilizing and possibly improving the lake's water quality, as the outflow joined the trophy fishing lake.

The sources of contaminants were the mill site with remnant concentrate, the drainage from the underground workings, and groundwater discharge from the tailings. These contaminant sources produced sediments heavily-laden with iron, zinc and copper. The contaminant loading had to remain in the lake sediments. This was compounded by the fact that the lake only had a 3-year retention period. Boojum's efforts centered on using a combination of biological polishing to sequester the contaminants in the treatment lake, and ARUM to bio-mineralize the contaminants in the narrow channel of the lake forming the outflow. A floating cover was initiated with brush cuttings placed on the ice during the winter. The cuttings extended the shores of the outflow channel.

Table 9.3 shows a mass balance between contaminants entering the lake and its sediments. Boojum quantified the tonnage of the major contaminants, copper, iron, sulfur and zinc. Three time periods were differentiated in preparing the mass balance. The first period was before any ecological engineering measures were

Table 9.4 Contaminant loading and mass balance for copper, iron, sulfur, and zinc in a biological polishing lake for three different time periods, based on different treatment regimes

	Cu			Fe			S			Zn		
Boomerang Lake Load in total tons												
	in	out	retain	in	out	retain	in	out	retain	in	out	retain
No Treatment (1987–1994)	2.6	0.7	1.9	355	9	345	461	239	221	101	22	79
Phosphate and Brush (1995–1999)	1.1	0.5	0.6	416	9	407	466	228	238	98	41	57
Magnesium (2000–2003)	0.8	0.6	0.2	314	11	303	339	244	95	88	47	41
Sediment Sink in total tons												
Sediment (1998)			2			468			na			51

implemented. The second period started when brush cuttings for the attachment of periphyton were added to the lake. At generally the same time, moss transplants were initiated, along with a onetime -phosphate fertilization. The third period began with the experimental suspension of metallic magnesium on barges.

The tonnages of contaminants which were retained in the sediments was estimated by determining the load of contaminants entering the lake (In) minus the load leaving the lake (Out). The historic or background loadings reported in Table 9.4 have been estimated from sediment grab samples obtained in 1990. The stratum at 5 cm was analyzed and used as background. The contaminant mass balance demonstrated that ecological measures not only led to the retention of contaminants, but these measures counteracted any deterioration of the water quality over the three decades since work began. Extensive documentation of the site can be found at the Laurentian Library as all the reports are available there under the title, South Bay.

The work carried out employed nearly all processes and tools discussed in this contribution. They are summarized here.

- Seasonal turnover of the shallow lake caused the iron-rich sediments to oxidize every year, driving the pH lower. A living moss cover over these sediments was initiated to prevent seasonal iron oxidation (See Chap. 5).
- Metallic magnesium was used as a means to relatively quickly increase the lake pH. The method worked, but could not be scaled enough to alter the lake pH (see Chap. 7).
- Approx. 140 tonnes of ground calcium phosphate mining wastes were added to the lake water and sediment to stimulate microbial, phytoplankton and moss cover growth (see Chap. 8).
- Cut brush was placed along the perimeter of the lake to add surface area for the establishment and growth of biofilms and periphyton and to jump start the process of terrestrialization (see Chaps. 6 and 7).
- Oxygen and water ingress into the tailings was slowed with 14 tonnes of calcium phosphate by forming a hardpan in the tailings, where annual ingress was suspected. The hardpan was expected to reduce seasonal water movement in the vadose zone and slow the groundwater movement (see Chap. 8).

Boojum terminated cooperation with Ministry of Mines and Northern Development in 2002 over disagreements about the ownership of the mine wastes and the associated responsibility. The site was returned to the crown decades ago with the full approval of the regulatory bodies of the time. To close the project, Boojum's client requested that a summary of the contaminant sinks and sources within the mine waste management area be prepared. The largest sink in the waste management area was the contaminated lake, with sediments heavily-laden with iron, zinc and copper. By that time, the ecological measures used on the lake had matured. The underwater meadow which developed from the transplanted moss, was completely covering the sediments. The narrow outflow of the lake, where brush was placed, had begun the process of 'landing in' (Fig. 9.21a).

The Ministry of Northern Development and Mines requested a comparison between the lake water quality monitoring completed by Boojum Research in 2002 and those measured by the ministry in 2013 and 2015. The unexpectedly good lake water quality in 2015, given the loadings from the contaminant sources, suggests that the ecological engineering processes continue to work, not only in the treatment lake (Fig. 9.21a), but in the tailings, as well (Fig. 9.21b). The tailings have a dense vegetation cover which might reduce infiltration of atmospheric precipitation. A summary report of the program, along with detailed descriptions of work accomplished is detailed Kalin (2003).

Boojum's ecological engineering tools will retain a large fraction of the metals within the mine waste management area and with that, a reasonable chance to solve many of mining's environmental challenges. All that is needed is a paradigm shift from thinking about mine wastes as toxic wastes to natural, weathering of uncovered, extreme ecosystems.

Fig. 9.21 The experimental lake and tailings 4 years after Boojum's departure. The lush green vegetation (both aquatic and terrestrial) is likely due to the application of CPMW to both lake and tailings. (**a**) Periphyton growth on cut brush in the biological polishing lake. (**b**) The growth of indigenous vegetation on the tailings after a CPMW addition. (Photographs taken in 2006 by the University of Windsor)

References

Baechler, F. (1997). Investigation of PERD piles at CBDC VJCPP plant. pp 57. https://zone.biblio. laurentian.ca/handle/10219/2923

Boojum Research Ltd. (1992). *La Mine Selbaie. Decommissioning with Ecological Engineering.* Boojum Research Ltd 88 pages. Retrieved from: https://zone.biblio.laurentian.ca/ handle/10219/3008

Boojum Research Ltd. (1994). The decommissioning of Buchans Unit and implementation of biological polishing. Final report / prepared for G. Neary, Buchans Unit, ASARCO. Retrieved from https://zone.biblio.laurentian.ca/handle/10219/2969

Boojum Research Ltd. (1997). *B-Zone Pit: Limnology 1993–1996 and the Fate of Arsenic and Nickel: Final Report. Produced for Cameco Corporation as CA105.* Retrieved from https:// zone.biblio.laurentian.ca/handle/10219/3037.

Boojum Research Ltd. (2002) Development of a pit lake and fate of contaminants 1992–2001, pp156 https://zone.biblio.laurentian.ca/bitstream/10219/2915/1/CA112.pdf

Cao, Y., & Kalin, M. (1999). Phytoplankton in mine waste water community structure, control factors and biological monitoring. *Natural Resources Canada, Biotechnology for Mining. Contract 23440–8.* Retrieved from https://zone.biblio.laurentian.ca/handle/10219/3015

Chen, X., Wright, J. J. V., Conca, J. J. L. J., & Peurrung, L. L. M. (1997). Evaluation of heavy metal remediation using mineral apatite. *Water, Air & Soil Pollution, 98,* 57–78. http://www. springerlink.com/index/l7l2p67804476680.pdf

Evangelou, V. P. (1995). *Pyrite oxidation and its control.* New York: CRC Press. pp 293 ISBN 9780849347320 – CAT# 4732.

Falkowski, P. G. (2000). Rationalizing elemental ratios in unicellular algae. *Journal of Phycology, 36*(1), 3–6.

Georgopoulou, Z. J., Fytas, K., Soto, H., & Evangelou, B. (1996). Feasibility and cost of creating an iron-phosphate coating on pyrrhotite to prevent oxidation. *Environmental Geology, 28*(2), 61–69.

Herbst, A., Patzelt, L., Schoebe, S., Schubert, H., & von Tümpling, W. (2019). Bioremediation approach using charophytes-preliminary laboratory and field studies of mine drainage water from the Mansfeld Region, Germany. *Environmental Science and Pollution Research.* https:// doi.org/10.1007/s11356-019-06552-6

Hutchinson, P. J., & Barta, L. S. (2000). Geophysical applications to solid waste analysis. In *The 16th international conference on solid waste technology and management, Philadelphia, PA, USA* (pp. 2–68).

Kalin, M. (2003). Closure with ecological engineering of a remote Cu/Zn concentrator: overview of 16 years R & D field program pp25. https://zone.biblio.laurentian.ca/handle/10219/2986

Kalin, M. (2004). Improving pore water quality in reactive tailings with phosphate mining wastes. *Proceedings of the Fifth International Symposium on Waste Processing and Recycling in Mineral and Metallurgical Industries, and the 43rd Annual Conference of Metallurgists of CIM,* Hamilton, Ontario, August 22–25, pp. 427–437.

Kalin, M. (1993). The application of ecological engineering to Selminco Summit. Final Report prepared for G. Landry, DEVCO. Retrieved from: https://zone.biblio.laurentian.ca/ handle/10219/2996

Kalin, M. (2009). Buchans: Ecological Engineering Treatment Assessment: Long-term performance evaluation and Site Visit Report. https://zone.biblio.laurentian.ca/handle/10219/2892

Kalin, M., & Pawlowski, J. (1994). Electromagnetic surveys in acid-generating waste management areas. In *Proceedings of the international symposium on extraction and processing for the treatment and minimization of wastes,* San Francisco, California, February 27–March 3, pp. 727–736.

Kalin, M., & Wheeler, W. N. (2013). Biological polishing of arsenic, nickel and zinc in an acidic lake and two alkaline lakes. In W. Geller, M. Schulze, R. Kleinman, & C. Wolkersdorfer (Eds.), *Acidic pit lakes: The legacy of coal and metal surface mines* (pp. 387–407). Springer.

Kalin, M., van Everdingen, R. O., & McCready, R. G. L. (1992). Ecological engineering- interpretation of hydrogeochemical observations in a sulphide tailings deposit. *CIM Bulletin, 85*(965), 64–67. Retrieved from https://www.researchgate.net/profile/Margarete_Kalin/publication/285730714_Ecological_engineering_-_Interpretation_of_hydrogeochemical_observations_in_a_sulphide_tailings_deposit/links/568e438a08aef987e56760cc/Ecological-engineering-Interpretation-of-hydr

Kalin, M., Fyson, A., & Smith, M. P. (1993). Heterotrophic bacteria and grass covers on fresh, base metal tailings. In *Proceedings of the Canadian land reclamation conference 18th annual meeting*, Lindsay, Ontario, August 11–13, pp. 81–88.

Kalin, M., Cao, Y., Smith, M. P., & Olaveson, M. M. (2001). Development of the phytoplankton community in a pit lake in relation to water quality changes. *Water Research, 35*(13), 3215–3225.

Kalin, M., Kiessig, G., & Küchler, A. (2002). Ecological water treatment processes for underground uranium mine water: Progress after three years of operating a constructed wetland. In B. J. Merkel, B. Planer-Friedrich, & C. Wolkersdorfer (Eds.), *Uranium in the aquatic environment* (pp. 587–596). Springer.

Küchler, A., Kiessig, G., & Kunze, C. (2006). Passive biological treatment systems of mine waters at WISMUT sites. In *Uranium in the environment* (pp. 329–340). Springer.

Marquardt, M., & Schubert, H. (2009). Photosynthetic characterization of *Chara vulgaris* in bioremediation ponds. *Charophytes, 2*, 1–8. https://www.researchgate.net/publication/229066985

WISUTEC. (2002). Jahresbericht 2002: Ergebnisse des Pilotversuches zur passiv/biologischen Behandlung multitasking on Grubenwasser der Grube Pohla-Tellhauser. Page 2.

Chapter 10
The Way Forward

Margarete Kalin-Seidenfaden ⓘ, Michael P. Sudbury, and Bryn Harris ⓘ

Abstract In the preceding chapters of this book, the authors have described and highlighted some of the problems facing our world with respect to Climate Change, and more especially those problems associated with mining and recovering the very minerals and metals that underpin our modern society, and which are needed to combat climate change. Extrapolating recent developments, this final chapter identifies the main problems that mining activities will both face and cause in the future, and discusses these recent developments, which it is hoped may help to solve them. The target here is to develop ideas for mining practices which not only do not contrast with, but rather, enhance and help to implement the **sustainable development goals (SDGs)** of the United Nations (UN, Department of Economic and Social Affairs, 2021).

Keywords Heavy metal pollution · Elements · Bioremediation · Sustainability goals · Ecological niche construction · Ecological engineering · Biogeochemical technologies · Hydrometallurgy · Metals extraction

In this chapter, the authors express their concerns after having spent their professional lives in this proud and essential industry. For its survival along with that of society and the planet, we see the need for a new direction, which we have presented here. The ecological engineering steps outlined in this book are only the beginning

M. Kalin-Seidenfaden (✉)
Boojum Research Ltd., Toronto, ON, Canada
e-mail: margarete.kalin@utoronto.ca

M. P. Sudbury
Michael P. Sudbury Consulting Services Inc., Oakville, ON, Canada
e-mail: msudbury@cogeco.ca

B. Harris
Dorine Road, Alexandria, Canada
e-mail: bryn@sutekh.org

M. Kalin-Seidenfaden, W. N. Wheeler (eds.), *Mine Wastes and Water, Ecological Engineering and Metals Extraction*,
https://doi.org/10.1007/978-3-030-84651-0_10

147

for remediation and management of existing tailings, but without more effective mineral extraction, the integration of ecological expertise into waste and water management, there will ultimately be too much stress put on the environment, and it will not be possible for metals extraction to keep up with the demands of society, especially when taken in conjunction with climate change and the electric future.

Water usage in the mining and metals extraction industry is on a collision course with society and agriculture. Some of our most important and critical metal ore deposits will be mined out in a couple of decades or even less, especially if the anticipated "electric future" is to become a reality. Acid mine drainage, an inevitable consequence of past and current mining practices, has been plaguing us for hundreds of years and will continue to haunt the industry for many centuries unless something is done.

Sustainability is possible but only with a considerable effort by the industry to change direction, which has been outlined with this contribution. The past (and present) methods are not working to achieve the global sustainability goals set by the **United Nations in 2015** (UN, Department of Economic and Social Affairs, 2021).

Generally, up to the present, mining, especially base metal mining, has focused only on one (or two) pay metals of interest, with the rest being regarded as gangue or nuisance material. This especially applies to iron, sulfur, calcium, magnesium and aluminum minerals. The favored mineral extraction method today is hydrometallurgical, rather than smelting. Hydrometallurgical methods can and should extract more metals out of the mined ore. This would improve the economics of the operations and at the same time reduce the global footprint of land consumption of mining wastes, reduce water consumption and hence move towards sustainability. In the past, environmental issues were largely ignored, it was simply the price to pay. The Sudbury Basin in Canada is a prime example, the area being akin to a moonscape in the 1950s now greened, but the drainage remains contaminated. Generally, scientific studies are funded, and the awareness of the public is raised by politics. The Sudbury Basin today is now unrecognizable from what it was, but it took many stakeholders working together to recover the local environment.

An 'outside the box' approach has been presented in the preceding chapters, outlining the ecological processes within mine waste and water management areas. Ecological engineering is viewed by some as a "do nothing option." While academia continues to clarify and understand the problem, it is now time to assist with the solution. A fundamental step in this direction is presented herein. This has to be further developed, possibly together with Nature Based Solutions. Many scientists recognize the potential of microbial activities in extreme ecosystems, and it has been clearly demonstrated that ecologically based processes effectively retain and fix the toxic metals within the wastes. Moreover, ecological engineering has a low-carbon intensity and, furthermore can help sequester carbon as biomass and carbonates. Improving water quality fosters biological productivity and results in direct carbon capture from the atmosphere. Therefore, ecological engineering tools have an intrinsic potential to support mining to achieve sustainability targets.

The journey of Boojum began in the late 1980s, at a time when biotechnology was becoming trendy. By serendipity, we followed the same path, albeit only the inhibition of oxidation required microbiological work. A wealth of information has been placed into the Laurentian library and can be used for teaching and as examples to improving and expanding the ecological engineering fundaments documented in the scientific literature. The mining companies participated in the work with the attitude that "it might work" and hoped that if it did, it would be accepted by the regulatory agencies, who unfortunately failed to see the value of the approach and the trust in natural processes. What needs to be done to overcome these concerns is outlined below:

Concern	Validity
Lack of Proven Technique for Biologically Inhibiting Sulphide Oxidation	Stimulated biofilms covering sulphidic minerals , showing longevity with considerable promise, to be economic and practical – research ongoing for 10 years with field tests
Biological Slow Down in Winter	Rock Piles have Great Thermal Mass - Algal Systems continue to function Under Ice (cf. Cod Feeding on Algae under Arctic Ice) Contaminant generation also slows down in winter
Lack of Solid Design Criteria - Capacity and Nutrient Needs	Need for Commercial Scale Piloting – utilizing existing demonstrations systems. Needs input and cooperation of mining company
Limited Capacity to Handle Droughts, Floods & Dissolved Solids Surges	Potentially better tolerance as engineered system, Optimization to extreme situations feasible – extreme ecosystems, desert or monsoonal areas)
Algal Filtration/Adsorbtion Capacity Scale up on a Volumetric Basis with Depth Limitation (light need)	Integration of engineering systems needed gap to be filled on Optimization processes.
Sediment Scale up Limited by Diffusion Across the Interface (Area Controlled)	Bio-stabilization in sediments needs more knowledge-research funds

The global consequences of inaction are evidenced already in the dramatic changes in climate, frequent floods, droughts, and water shortages, all factors affecting mining operations intensely. It can only be hoped that in the future the deep chasms between science and engineering will narrow as application will narrow, as many disciplines are needed to address mining waste and water.

In conclusion, then, we believe that the mining industry needs a paradigm shift in the way it operates. It has to be aware of the consequences of mining, in terms of sustainability (i.e. making the most of finite resources), of land and water usage, and dealing with legacy mining sites. Improved hydrometallurgical extraction, recovering as much of the valuable components of the ore mined, can play a significant part in this respect. Hydrometallurgy is also suited for remediating a number of old mining sites and tailings facilities, where metals can be recovered at the same time. However, in the overall scheme of things, we believe that ecological engineering will ultimately have the biggest role to play, particularly in allowing historical sites to be "returned to nature." There are ample data presented in this book to demonstrate the effectiveness of this approach. It is far superior to treating AMD in perpetuity with lime. Mining and metals extraction will continue to be essential to human progress, but current practices have to be modified if we are to protect our environment and show the appropriate stewardship of our home, the Blue Marble, planet Earth.

Reference

UN, Department of Economic and Social Affairs. (2021). *Sustainability goals.* https://sdgs. un.org/goals

Related Reading

Chapter 1: Introduction and Weathering

Moon, S., Rosenblum, F., Tan, Y., Waters, K. E., & Finch, J. A. (2020). Transition of sulphide self-heating from stage A to stage B. *Minerals, 10*(12), 1133. https://doi.org/10.3390/min10121133

Rosenblum, F., Nesset, J. E., Moon, S., Finch, J. A., & Waters, K. E. (2017). Reducing the self-heating of sulphides by chemical treatment with lignosulfonates. *Minerals Engineering, 107*, 78–80.

Chapter 2: Dimensions of Global Mining Waste Generation and Water Use

Herrera-Leon, S., Cruz, C., Kraslawski, A., & Cisternas, L. A. (2019). Current situation and major challenges of desalination in Chile. *Desalination. Water Treat, 171*, 93–104. https://doi.org/10.5004/dwt.2019.24863

Ives, M. (2013). Boom in mining rare earths poses mounting toxic risks. *Yale Environment, 360*, 28. https://e360.yale.edu/features/boom_in_mining_rare_earths_poses_mounting_toxic_risks

Knops, F., Kahne, E., Mata, M. G. D. L., & Fajardo, C. M. (2013). Seawater desalination off the Chilean coast for water supply to the mining industry. *Desalination and Water Treatment, 51*(1–3), 11–18.

Mudd, G. M., & Jowitt, S. M. (2016). From mineral resources to sustainable mining – The key trends to unlock the holy grail. In *Proceedings of the third AusIMM international Geometallurgy Conference (GeoMet) 2016* (pp. 37–54).

Northey, S. A., Mudd, G. M., Werner, T. T., Jowitt, S. M., Haque, N., Yellishetty, M., & Weng, Z. (2017). The exposure of global base metal resources to water criticality, scarcity and climate change. *Global Environmental Change, 44*, 109–124.

Tost, M., Bayer, B., Hitch, M., Lutter, S., Moser, P., & Feiel, S. (2018). Metal mining's environmental pressures: A review and updated estimates on CO2 emissions, water use, and land requirements. *Sustainability, 10*(8), 2881. https://doi.org/10.3390/su10082881

M. Kalin-Seidenfaden, W. N. Wheeler (eds.), *Mine Wastes and Water, Ecological Engineering and Metals Extraction*, https://doi.org/10.1007/978-3-030-84651-0

Chapter 3: Toward a Sustainable Metals Extraction Technology

Sovacool, B. K., Ali, S. H., Bazilian, M., Radley, B., Nemery, B., Okatz, J., & Mulvaney, D. (2020). Sustainable minerals and metals for a low-carbon future. *Science, 367*(6473), 30–33. https://doi.org/10.1126/science.aaz6003

Chapter 4: Waste Management: A Brief History and the Present State

International Atomic Energy Association. (2020). https://inis.iaea.org/search/search.aspx?orig_q=author:%22Kalin,%20M.%2. Retrieved on 1 Nov 2021.

Lottermoser, B. G. (2010). *Mine wastes: Characterization, treatment and environmental impacts.* Springer. https://doi.org/10.1007/978-3-642-12419-8

Xu, D. M., Zhan, C. L., Liu, H. X., & Lin, H. Z. (2019). A critical review on environmental implications, recycling strategies, and ecological remediation for mine tailings. *Environmental Science and Pollution Research, 26*(35), 35657–35669. https://doi.org/10.1007/s11356-019-06555-3

Chapter 5: Constructed Wetlands and the Ecology of Extreme Ecosystems

Donati, E. R., Sani, R. K., Goh, K. M., & Chan, K. G. (2019). Recent advances in bioremediation/biodegradation by extreme microorganisms. *Frontiers in Microbiology, 10*, 1851. https://doi.org/10.3389/fmicb.2019.01851

Masi, F., Rizzo, A., & Regelsberger, M. (2018). The role of constructed wetlands in a new circular economy, resource oriented, and ecosystem services paradigm. *Journal of Environmental Management, 216*, 275–284. https://doi.org/10.1016/j.jenvman.2017.11.086

Salmon, S. U., Hipsey, M. R., Wake, G. W., Ivey, G. N., & Oldham, C. E. (2017). Quantifying lake water quality evolution: Coupled geochemistry, hydrodynamics, and aquatic ecology in an acidic pit lake. *Environmental Science & Technology, 51*(17), 9864–9875. https://doi.org/10.1021/acs.est.7b01432

Sharma, R., Vymazal, J., & Malaviya, P. (2021). Application of floating treatment wetlands for stormwater runoff: A critical review of the recent developments with emphasis on heavy metals and nutrient removal. *Science of the Total Environment, 146044.*

Chapter 6: Ecological Engineering Tools in Extreme Ecosystems

Chen, H., Xiao, T., Ning, Z., Li, Q., Xiao, E., Liu, Y., … Lu, F. (2020). In-situ remediation of acid mine drainage from abandoned coal mine by filed pilot-scale passive treatment system: Performance and response of microbial communities to low pH and elevated Fe. *Bioresource Technology, 317*, 123985. https://doi.org/10.1016/j.biortech.2020.123985

Sekarjannah, F., Mansur, I., & Abidin, Z. (2021). Selection of organic materials potentially used to enhance bioremediation of acid mine drainage. *Journal of Degraded and Mining Lands Management, 8*(3), 2779–2789. https://doi.org/10.15243/jdmlm.2021.083.2779

Chapter 7: Biological Polishing Tool: Element Removal in the Water Column

Blindlow, I., Carlsson, M., & van de Weyer, K. (2021). Re-establishment techniques and transplantations of charophytes to support threatened species. *Plants, 10*, 1830–1863.

Dupraz, C., & Visscher, P. T. (2005a). Microbial lithification in marine stromatolites and hypersaline mats. *Trends in Microbiology, 13*(9), 429–438. https://doi.org/10.1016/j.tim.2005.07.008

Nixdorf, B., Fyson, A., & Krumbeck, H. (2001). Plant life in extremely acidic waters. *Environmental and Experimental Botany, 46*(3), 203–211.

Pełechaty, M., Ossowska, J., Pukacz, A., Apolinarska, K., & Siepak, M. (2015). Site-dependent species composition, structure and environmental conditions of *Chara tomentosa* L. meadows, western Poland. *Aquatic Botany, 120*, 92–100.

Schneider, S. C., García, A., Martín-Closas, C., & Chivas, A. R. (2015). The role of charophytes (Charales) in past and present environments: An overview. *Aquatic Botany, 120*, 2–6. https://doi.org/10.1016/j.aquabot.2014.10.001

Chapter 8: The Biofilm Generation Tool for the Reduction of Sulfate Oxidation

Al Qabany, A., Soga, K., & Santamarina, C. (2012). Factors affecting efficiency of microbially induced calcite precipitation. *Journal of Geotechnical and Geoenvironmental Engineering, 138*(8), 992–1001.

Dupraz, C., & Visscher, P. T. (2005b). Microbial lithification in marine stromatolites and hypersaline mats. *Trends in Microbiology, 3*(9), 429–438.

Idrissi, H., Taha, Y., Elghali, A., El Khessaimi, Y., Aboulayt, A., Amalik, J., … Benzaazoua, M. (2021). Sustainable use of phosphate waste rocks: From characterization to potential applications. *Materials Chemistry and Physics, 260*, 124119. https://doi.org/10.1016/j.matchemphys.2020.124119

Laland, K., Odling-Smee, J., & Feldman, M. (2000). Niche construction, biological evolution, and cultural change. *The Behavioral and Brain Sciences, 23*(1), 131–146. https://doi.org/10.1017/S0140525X00002417

Little, B., Lee, J., & Ray, R. (2007). A review of 'green' strategies to prevent or mitigate microbiologically influenced corrosion. *Biofouling, 23*(2), 87–97. https://doi.org/10.1080/08927010601151782

Madsen, E. L. (2011). Microorganisms and their roles in fundamental biogeochemical cycles. *Current Opinion in Biotechnology, 22*(3), 456–464. https://doi.org/10.1016/j.copbio.2011.01.008

Rahman, M. M., Hora, R. N., Ahenkorah, I., Beecham, S., Karim, M. R., & Iqbal, A. (2020). State-of-the-art review of microbial-induced calcite precipitation and its sustainability in engineering applications. *Sustainability, 12*(15), 6281.

Zhang, R., Duan, J., Xu, D., Xia, J., Muñoz, J. A., & Sand, W. (2021). Bioleaching and biocorrosion: Advances in interfacial processes. *Frontiers in Microbiology, 12*, 653029.

Chapter 9: R&D Field Applications

Boojum Research Ltd. (1994). The decommissioning of Buchans Unit and implementation of biological polishing. Final report / prepared for G. Neary, Buchans Unit, ASARCO. Retrieved from https://zone.biblio.laurentian.ca/handle/10219/2969.

Kalin, M. (2009). Buchans: Ecological Engineering Treatment Assessment: Long-term performance evaluation and Site Visit Report. https://zone.biblio.laurentian.ca/handle/10219/2892.

Kalin, M. (1993). The application of ecological engineering to Selminco Summit. Final Report prepared for G. Landry, DEVCO. Retrieved from: https://zone.biblio.laurentian.ca/handle/10219/2996.

Chapter 10: The Way Forward

Daily, G. C., & Walker, B. H. (2000). Seeking the great transition. *Nature, 403*, 243–245. https://doi.org/10.1038/35002194

Dold, B. (2020). Sourcing of critical elements and industrial minerals from mine wastes – The final evolutionary step back to sustainability of humankind. *Journal of Geochemical Exploration., 219*, 106638. https://doi.org/10.1016/j.gexplo.2020.106638

Grande, J. A., Santisteban, M., De La Torre, M. L., Fortes, J. C., De Miguel, E., Curiel, J., & Biosca, B. (2018). The paradigm of circular mining in the world: The Iberian Pyrite Belt as a potential scenario of interaction. *Environmental Earth Sciences, 77*(10), 1–6.

International Institute of Environment and Development (IIED). (2012). *Mining, minerals and sustainable sevelopment.* https://www.iied.org/mining-minerals-sustainable-development-mmsd. Retrieved on 1 Nov 2021.

Marxsen, J., & (Ed.). (2016). *Climate change and microbial ecology.* Caister Academic Press. https://doi.org/10.21775/9781913652579

Ndlovu, S., Simate, G. S., & Matinde, E. (2017). *Waste production and utilization in the metal extraction industry.* CRC Press.

Simate, G. S., Ndlovu, S., & (Eds.). (2021). *Acid mine drainage: From waste to resources.* CRC Press.

Printed in the United States
by Baker & Taylor Publisher Services